The City as Action

In constructing the urban as a set of interconnected actions, this book presents a less travelled route to understanding the city. It leads to a fresh perspective on several issues central to urban theory, including the uniqueness of a city alongside practices it shares with other urban places.

This book presents an innovative theoretical contribution to the field of urban studies, bridging the gap between western centric scholarship and perspectives from the global South. It offers conceptually rich insights, combining notions of cities as organisms, and references to postcolonial urban studies, with insights around aspirations, capabilities, agency, and social identity. It develops concepts, like the Proximity Principle, that help explain the experience of a city.

This conceptualization of the city as a process should interest all who are sensitive to cities, whether they study them in academia or simply develop close associations with specific urban places.

Narendar Pani is Professor and Dean, School of Social Sciences, National Institute of Advanced Studies, India.

Global Urban Studies

Providing cutting edge interdisciplinary research on spatial, political, cultural and economic processes and issues in urban areas across the US and the world, books in this series examine the global processes that impact and unite urban areas. The organizing theme of the book series is the reality that behavior within and between cities and urban regions must be understood in a larger domestic and international context. An explicitly comparative approach to understanding urban issues and problems allows scholars and students to consider and analyse new ways in which urban areas across different societies and within the same society interact with each other and address a common set of challenges or issues. Books in the series cover topics which are common to urban areas globally, yet illustrate the similarities and differences in conditions, approaches, and solutions across the world, such as environment/brownfields, sustainability, health, economic development, culture, governance and national security. In short, the Global Urban Studies book series takes an interdisciplinary approach to emergent urban issues using a global or comparative perspective.

Metropolis, Money and Markets
Brazilian Urban Financialization in Times of Re-emerging Global Finance
Jeroen Klink

Animals in the City
Edited by Laura A. Reese

Twin Cities across Five Continents
Interactions and Tensions on Urban Borders
Edited by Ekaterina Mikhailova and John Garrard

The City as Action
Retheorizing Urban Studies
Narendar Pani

The City as Action
Retheorizing Urban Studies

Narendar Pani

Routledge
Taylor & Francis Group

LONDON AND NEW YORK

First published 2022
by Routledge
4 Park Square, Milton Park, Abingdon, Oxon OX14 4RN

and by Routledge
605 Third Avenue, New York, NY 10158

Routledge is an imprint of the Taylor & Francis Group, an informa business

British Library Cataloguing-in-Publication Data
A catalogue record for this book is available from the British Library

Library of Congress Cataloging-in-Publication Data
A catalog record has been requested for this book

ISBN: 978-1-032-05267-0 (hbk)
ISBN: 978-1-032-05268-7 (pbk)
ISBN: 978-1-003-19679-2 (ebk)

DOI: 10.4324/9781003196792

Typeset in Bembo
by Taylor & Francis Books

For Jamuna, for everything

Contents

.

Foreword

Urban studies is an amorphous field, approached by analysts from all the social science disciplines as well as architecture and engineering. Scholars from each of these constituent disciplines bring their own biases and definitions to the meaning of city and urban. For geographers the city is defined spatially; for sociologists human interactions that form structures and coalesce or block community ties constitute the key to understanding. Urban planners focus on public policy, anthropologists on culture, while political scientists analyze governance, and economists highlight markets. Architects, designers, and engineers see the built environment. Narendar Pani in this volume takes on the herculean task of combining these differing points of view into an overarching theory of action. He distinguishes between the urban and the city — the latter a static notion if defined only by territorial boundaries but a flexible concept if seen in terms of processes. Pani's identification of the city with agglomeration escapes the "territorial trap" through regarding the space of the city as variable in relationship to differing types of agglomeration and points to the way, especially within the global South, that city boundaries can encompass rural enclaves even while these enclaves are transformed by urbanization.

Adopting Scott and Storper's (2015) conception of the urban, Pani refers to the process of agglomeration cum polarization as its defining characteristic. This theorization, a descendant of Trotsky's theory of combined and uneven development, emphasizes that urbanization rather than being a homogeneous process creates difference along with commonality. The various communities formed within the larger agglomeration perform acts of both inclusion and exclusion, incorporating shared beliefs and hostilities. The process of agglomeration is constantly mutating, responding to actions within and outside territorial boundaries. These actions include intentional choices but also are responses to "what happens", i.e. responses to situations that constrain options. The city is an agglomeration created by the actions of its inhabitants and also by the actions of others who may occupy spaces far away.

In an unusual approach, Pani illustrates his argument concerning the effects of agglomeration by examining the range of choice available to his heroine Wimoa ("Woman In the Midst of Agglomeration"). Centering his analysis on a woman, he avoids many of the preconceptions that bias other inquiries. He calls her his lodestar, as she migrates from a rural village to a city. Consequently, the actions he discusses are not simply abstractions but rather the decisions made by or forced upon her within the spatial and social structures of the city. Wimoa, a woman near the bottom of the social hierarchy, has migrated to the city to join her husband and must find employment and social support within her new surroundings. She both intends certain acts – seeking a job, choosing friends, buying food, etc. – and experiences actions, intended and unintended by others, that affect her. Thus, the act of investment by a multinational firm headquartered in a distant global city determines the kinds of jobs and pay levels within Wimoa's reach. The existence of religious and ethnic communities within her city constrains her social network and may either protect or endanger her. Moreover, individual behaviors may aggregate into group actions that go beyond the intentions of its members. It is the complicated interactions of individuals and groups within and beyond a city that constitute the urban process.

Pani uses his methodological focus on action as the unit of analysis to examine the economic, religious, spatial, and social nodes and networks that we identify as constituting a city. This allows him to trace connections between the command-and-control centers of global capital, technological change, and the growth of places offering access to skilled and unskilled populations that enable least-cost production. In turn, the gathering of individuals who make up labor pools gives rise to new social affinities and local politics. Then, these outcomes may affect the calculations of investors who see labor costs rising in the original location of production and may decide to move their call center or factory elsewhere, resulting in unemployment for those attracted to the city by its job opportunities. At the same time, the initial infusion of resources into these areas might have improved educational levels and stimulated the development of a more highly skilled labor force capable of starting indigenous firms and providing expertise based on advanced technologies, as occurred, for instance, in Bangaluru (formerly Bangalore).

For Pani spatial boundaries, economic entities, ethnic groupings, and governing institutions do not fix the character of the city. Although we use the terms New Delhi or Singapore to refer to a unique territory defined by an overarching government, Pani regards each as a collection of places not a single place. Sports teams may constitute a primary way in which residents identify with the city as a whole, but the entity labeled a city may incorporate business establishments governed by decisions made elsewhere and include former villages that maintain their old internal hierarchies and customs. In fact, these rural enclaves may continue to rely on agricultural production as their economic base. The Marxian formulation that the relations of production determine social outcomes and that the state is the executive committee of the bourgeoisie does not explain urban development even while it may portray individual opportunities and livelihoods. Neither does the rubric of planetary urbanization (Brenner & Schmid, 2015), which Pani argues fails to distinguish between the urban and rural or account for ongoing affiliations with villages of origin to which migrants may commute on a seasonal basis.

The City as Action offers a methodology for studying cities that is premised on indeterminacy. It identifies sources of urban conflict, but Pani, unlike the critical urban theorists who blame capitalism and neoliberalism for urban ills, does not take sides even while he identifies injustices. His approach is akin to actor-network theories, where the social world exists in a constantly changing set of relationships. For those who see their mission as not simply interpreting the world but changing it, this book is daunting in pointing to the unintended consequences of action but also encouraging in stressing the ongoing importance of the local. Wimoa's circumstances offer a starting point for reformers. Providing her with a framework that protects women from predators, offers them greater education and training, and improves both the housing and the physical infrastructure of their neighborhoods would go far in transforming what happens to Wimoa into her greater power to make things happen. Professor Pani, in giving us a methodology for delineating urban processes, also is offering us an approach to effective strategies for improving urban outcomes.

Susan S. Fainstein
Graduate School of Design
Harvard University

References

Brenner, N. & Schmid, C., 2015. Towards a new epistemology of the urban?. *City*, 19 (2–3), pp. 151–182.

Scott, A. J. & Storper, M., 2015. The nature of cities: The scope and limits of urban theory. *International Journal of Urban and Regional Research*, 39(1), pp. 1–15.

Preface

Having lived much of my life in the south Indian city of what was once Bangalore, and is now Bengaluru, I have been an unintentional participant observer in the transformation of a rather laid-back city into a much more hurried megacity. My recollection of that transformation is almost entirely in terms of what happened in the city: the actions of its inhabitants, sometimes intentional and at other times due to factors beyond their control. In school in the 1960s I was witness to Bengaluru's first wave of emigration, as the Anglo-Indian community preferred foreign shores, especially Australia. My school, which was set up for Anglo-Indians in 1854, suddenly found their numbers dwindling and by the next decade there were hardly any around. The 1970s were marked by the beginning of Bengaluru's rapid growth. The decade saw fashion in the West change to include the clothes of the working class, resulting in global brands looking for cheaper places for garment manufacture elsewhere in the world. Bengaluru emerged as one such place, attracting young workers from nearby villages. At the same time, as the city spread to absorb neighbouring villages, these villages-within-cities became a natural destination for former rural workers who the garment industry had attracted to the city. The workers brought Kannada, the language of their villages and the larger Karnataka state in which Bengaluru is situated, to the vicinity of the Tamil-speaking sections of what was then Bangalore cantonment. This added, in the 1980s, the odd language riot to the occasional Hindu-Muslim riots that the city had already been witness to. This was also a time of English-speaking, technically trained engineers in the city seeking greener pastures in the West, generating a second wave of emigration. The coming of the twenty-first century and its advances in communication technology ensured that some of this trained manpower could be tapped by global firms, even as they remained situated in Bengaluru. While these jobs should have had a dampening effect on the need to migrate, the proximity to Western products and work conditions only increased the aspirations to move to the West.

As substantial as the influence of these actions has been on my recollection of Bengaluru's transformation into a megacity, conventional wisdom provides them little place in discussions about the city. When, in the 1980s, I first began researching and writing about Bengaluru and the urban in general, it was clear that the official view of the city was in direct conflict with the actions of its inhabitants. In that decade official documents still had a place for a Green Belt that would provide the lung space for the city. This plan required policy to override the overwhelming desire of the city's inhabitants to tap into the rising real estate values provided by rapid growth. Actions of the inhabitants soon proved to be more powerful than the words of policy makers. The twenty-first century has seen more diverse influences on policies for the city. In the early years of this century the vision for the city was to be determined by task forces, which consisted primarily of representatives of the information technology industry and those who thought like them. This vision too did not always gel with the actions of the city's inhabitants.

The divergence between perceptions of the city and the actions of its inhabitants extends beyond Bengaluru. Cities of the global South are replete with examples of the actions of their inhabitants running counter to norms derived from official perceptions of how the cities should be. The divergence is usually brushed under a variety of carpets. It is sometimes seen as evidence of the failure of law enforcement, and at other times a sign of how prone the citizens are to corruption. Academia has been less dismissive of these processes and has tended to develop concepts that explain specific forms of deviation of the actions of urban inhabitants from the norms of the city, such as occupancy urbanism. But it has stopped well short of seeing the city itself as a set of actions. This book builds a case for cities to be seen as a set of interconnected actions.

In bringing actions to the centre-stage of the story of urbanization, this book has had to not just relate to the many insights urban theory has to offer, but also borrow from other disciplines. The conceptualization of action has relied quite heavily on philosophical explorations, just as the Proximity Principle builds on the political philosophy of MK Gandhi. Broadening the canvas in which the urban is to be understood has demanded a method that is not only interdisciplinary but also open to multiple perspectives, moving from the particular to the general and back. The method allows the book to relate to theories of the urban developed in the global North as well as the discourse around urbanization in the global South.

This has been a long drawn out effort, which has benefitted from numerous conversations with scholars from different parts of the world, though many of them hold views that are quite at variance with what this book advocates. I owe a special debt of gratitude to Michael Goldman, Solomon Benjamin, Gautam Bhan, Patrick Heller, Ashutosh Varshney, Anthony Levitas, Sai Balakrishnan, Ian Scoones, Gordon McGranahan, Saurabh Arora, Amita Bhide, and Darshini Mahadevia. The book has also benefited from discussions with my students over the years. Throughout the extended period over which the ideas of this book have been developed, Jamuna has been a major source of intellectually critical and emotionally warm support.

1   In the midst of the urban

As the smog settled on a cold winter night in Delhi, a woman entering the city for the first time found herself desperately wishing she were not alone. When she had been put on that bus by her brother on the highway near her village, others around her had not thought too much about her traveling alone. When the bus broke down on the way she had been left alone to sit on a rock away from the others. By the time the bus restarted and the sun had begun to set, her being alone had become a topic for whispers in the bus. She knew her husband would be waiting for her at the bus stop in the city. He had told her about how unsafe the city was and would surely not leave her alone in it at night. Yet, in a corner of her mind, doubt began to grow. He had just come to the city a few months ago and may not still know it well enough. Would his new job give him the time to wait hours for her? As the bus slowly made its way through the thick smog she was already missing her year-old daughter who she had to leave behind with her parents in the village. Her husband had told her they would not have the time or the money to take care of their daughter in the city. She wondered again what it was that had made her undertake that journey. The simple answer was that her husband told her to, but what was it that had convinced her husband that this was the right thing to do? What was it that had convinced so many like her husband to move to a city at great personal cost?

On the flight to Riyadh in Saudi Arabia a woman software professional found herself wondering about her journey. She would think about the hijab in her cabin baggage, the veil she would have to wear when she went out in the city she was going to live in. Not being a Muslim she had never had to wear a hijab before. Having been an activist in her college days fighting for a variety of women's rights, she was not quite sure what it would be like living in a country where the right of a woman to drive a car had been a critical issue. She was able to get the job in the software centre only because it was an all-women workplace, with limited scope for dealing face-to-face with men. Riyadh was a city which celebrated so many of the values she was contemptuous of, yet she felt quite liberated by the opportunity to work in it. Coming out of an abusive marriage had not been easy. Seeking a divorce had been frowned upon in her still-conservative Indian city. She had felt a desperate need to move out, and Riyadh had provided

DOI: 10.4324/9781003196792-1

that opportunity. On the long flight her analytical mind kept going back to the relationship between the individual and the city. What was it about cities that brought hundreds of thousands of people together, from very different cultures, and still gave them a degree of autonomy to be themselves?

The woman in the queue outside immigration in an airport in Boston would have loved to have had a little more autonomy in her life. She had been married for two years to a husband she had barely met. In their search for an ideal matrimonial match, her Indian family had found a distant relative with a Green Card in the United States. The costs of a wedding good enough for a son-in-law employed in the United States had put a huge financial strain on her parents. They bore the burden, happy in the conviction that the future of their daughter was now secure. The groom had flown in to get married and left soon after the ceremony was over. She had not heard from him for months after the wedding. After a year of waiting for him to call her to the United States, her relatives had begun to get restless. They activated their social network in Boston to get her legal husband to invite her to his home. Another year later he had sent the required documents to her father and she was now entering the United States for the first time. She had heard of it being the land of freedom, but the thought of living in a strange man's house just because he was her husband made her quite apprehensive. If things went wrong, what was the institutional and other support she could fall back on? She fought hard to get her hopes to overcome her fears, thinking about how a person could develop social networks in an alien city.

These three little tales are not stories of real individuals but they are real in a way that only an abstraction can be: each of the women in them represents the experiences of several other women entering a new urban situation. If these composite women, or their intellectually inclined friends, were to fall back on academia to help them explore the urban condition that awaits them they could easily be drowned in facts. They would find no dearth of empirical evidence of an "urban turn". The academic debate on urbanization has generated substantial empirical detail, ranging from the celebration of the world crossing the point where a majority of its population lives in variously defined urban centres (Gleeson, 2012), to the informality that marks the functioning of megacities in the developing countries (Mngutyo & Jonathan, 2015); from the rapid urbanization in parts of the developing world (Cohen, 2004), to the influence of global urban command and control centres on the rest of the world (Taylor & Csomós, 2012). If the women were not to be satisfied with mere facts, though, and sought to know how and why cities behave the way they do, they would be met by a variety of not-always-consistent theories and piecemeal responses with a sense of urgency to them.

Policy makers in the parts of the emerging and the less developed worlds that have contributed the most to urbanization have scrambled to meet the challenges of their megacities (Huang, et al., 2016). Non-Government Organizations have, with varying degrees of success and conviction, sought to address the pain that almost inevitably results from rapid social and demographic transitions, especially in economically challenged environments (Fisher, 1997). Academia has responded with a spurt in urban studies that includes drawing the experience of urbanization in the global South into the larger debates on the urban (Roy & Ong, 2011). These efforts to develop a consistent theoretical explanation for what some see as the "urban age" have resulted in several conceptual innovations. World cities theorists have provided both methods for drawing up hierarchies of cities as well as, more promisingly, tools to capture the urban dynamics of globalization. Others have drawn inputs from Critical Social Theory to develop theories of the urban. And yet in this rich and growing body of urban literature it is difficult to miss a sense that the task of developing a comprehensive theory of the urban is far from complete. No school of urban thought has the widespread support that the Chicago School once enjoyed. Instead, as Neil Brenner noted in the second decade of the twenty-first century,

> Even a cursory examination of recent works of urban theory reveals that foundational disagreements prevail regarding nearly every imaginable issue – from the conceptualization of *what* urbanists are (or should be) trying to study to the justification for *why* they are (or should be) doing so and the elaboration of *how* best to pursue their agendas.
>
> (Brenner, 2013, p. 92)

There is thus no agreement on even whether the travails of the three composite women I started this exploration with should be a part of urban studies, let alone how best to understand their urban condition.

I enter this contentious terrain not to add another theory to it. There is no dearth of theories to explain specific urban processes, whether it is the circuits of globalization or the nature of agglomeration. I enter it to build on a conviction that the confusion over theorising the urban is, in large part, due to urbanists seeking theories to address what should be a search for an adequate methodology to study the urban. This introductory chapter will make the case for this conviction. The book will proceed to develop a method to understand the urban; a method that will develop the concepts that are needed to analyse that process and its relationship with the city.

The development of this alternative method of visualizing a city will begin with an analytical exploration of the urban that would take it to the vast multitude of relationships that are found in that set of processes. In laying out these relationships, and their relationship with each other, the book will use as a lodestar usually neglected actors in the urban story. The three tales I started with involved composite women who represented a selection of features of other women in the process of moving into an urban experience. I now go a step further to create an even broader composite woman who has elements chosen from the women of the three stories and those of many others. This Woman In the Midst of Agglomeration, Wimoa for short, acts as a lodestar for the exploration of the urban. She is much closer to the bottom of the economic hierarchy than to the top, and is more likely to represent the concerns of the economically and socially underprivileged, in addition to that of the gender that is widely discriminated against. She has also been placed in a megacity of the emerging and less developed worlds to capture an area of rapid urbanization that is not always at the top of the list of concerns of urban theory.

Adopting Wimoa as the lodestar for this book is not without its challenges. For one, her voice is best captured through those of women. In bringing together these voices in an earlier work on women at the threshold of globalization, I had the benefit of a woman co-author with a remarkable ability to strike a rapport with young women workers (Pani & Singh, 2012). The theoretical nature of this book does not provide much room for a similar recording of the voices of individual women, nor does it have the benefit of a woman as its author. The difficulties of developing this book around Wimoa extend beyond her gender. She has been placed at a point in the socio-economic hierarchy where she has to deal with economic and social discrimination. In a country like India the earning of her household would place her at around the poverty line, that is, she would be among the poor but not so poor that she cannot afford the basic costs of migrating to the bottom of the economic hierarchy in the city. The poverty that drives humans to live in the inhuman conditions – which is all that they can afford – in the periphery of a city in the global South, can only be fully understood by those who have experienced it. This author has only been a witness to this level of poverty and has not directly experienced it. Wimoa's movement from a village to the lower economic strata of a city would also involve her personally experiencing various forms of social discrimination, whether it is based on her caste, tribe, religion, language, ethnicity, or any other identity she may have. This experience is often too complex, and personally damaging, to be fully understood by an outsider. To take an extreme case, those who have lived in close proximity to the vestiges of untouchability in India undergo an experience that others can't completely fathom. Here again, I have been a witness to, but have not personally experienced, extreme social discrimination.

If, despite the lack of personal experience of some of the characteristics I have attributed to the fictional Wimoa, I have chosen the impact on her to guide the course of this book, it is because of the need to bring gender, as well as economic and social discrimination, closer to the core of urban theory. The fact that the woman has a prominent place in the process of urbanization is quite evident in the global South. Operating within prevailing patriarchal norms, women face the onerous task of maintaining a household in the midst of the uncertainties faced by a family that has just moved to the city. Oddly enough, the domestic workload of women can be even greater in processes of migration where the man sets out to the city alone in search of a job. He uses his social capital to stay at the home of a relative or friend in the city, adding to the burden on the woman maintaining that household. And that woman is often doing so in addition to holding a regular job outside the home (Pani & Singh, 2012). What is more, the nature of gender relations can also alter the course of the process of urbanization. In parts of India characterised by a substantial number of rural workers not being employed for even six months in a year, these workers typically do not have the resources to migrate permanently to expensive cities (Haque, 2022). They are then forced to migrate to the city for short periods in order to maintain their households in the village. The short-term migration usually has a strong gender element to it. The groups of workers who are brought together from villages to work in cities in very difficult shared living conditions, often have little room for women. These large all-male groups of young workers have an impact on the nature of the city they migrate to. They contribute to the urban sprawl, even as their continued loyalty to the village hampers the development of a city identity. And the absence of women in the group can influence the response, particularly that of its adolescent members, to other women in the city.

The nature of a city in the global South can also be affected by socio-economic discrimination in the village. Extreme forms of social discrimination, such as that faced by the Dalits in rural India, can add momentum to the economic pressures that drive them to the city. The degree of anonymity the city provides can buffer some of the extreme social discrimination that is a part of their everyday life in the village. But this process is by no means complete. Some of the discrimination can continue in social relations in the city as well. The interaction between the socio-economic discrimination in the village and that of the city contributes to the emergence of unique patterns of socio-economic inequality in each city; patterns that cannot be brushed aside when exploring the urban. These concerns are shared, in a way, by those, like Ananya Roy, who are interested in the "project of postcolonial urbanism and how the study of cities can be enriched through a renewed engagement with postcolonial studies" (Roy, 2011, p. 307).

Wimoa, as a lodestar, allows for not just women's interests but also patterns of socio-economic discrimination in the processes of urbanization to be brought into the urban theoretical construct. My not sharing Wimoa's gender, or her experience of socio-economic discrimination, could contribute to missing some aspects of her involvement in the process of urbanization, even as my lived experience of urbanization in the global South may add the odd fresh perspective to the study of the urban and of the city. This book's efforts to explore the urban through Wimoa's eyes may be less than perfect, and just one of the diverse ways in which the urban, and its relationship with the city, can be visualised, but it would help make the method of action that it develops available to even the usually neglected sections of the city.

The challenges of diversity

A major part of the blame for the unfinished theoretical tasks of urban studies must be placed at the door of diversity. There are far too many factors pulling urbanization in very different directions across the world to enable an easy fit into a comprehensive urban theory. The diversity begins at something as basic as the terrain of individual cities. Riyadh, in the middle of the Saudi Arabian desert, where petrol is not always much more expensive than water, can have its population spread out over large sprawling tracts of land. In contrast, Mumbai, with its severe constraints of both land and transportation, provides a very different picture of an urban sprawl. Again, Riyadh may have malls comparable in size and design to those in, say, Washington DC, but the experience of being in one is very different in the two cities, especially for women. These cultural differences are, in fact, quite widespread over both time and place. There was a time when Hindus, in pursuit of other-world interests, left their homes to go to Varanasi on the banks of the holy river Ganges to die, in the hope that it would help them in their afterlife (Gesler & Pierce, 2000). In other cultures people can go to a city in pursuit of this-world interests, such as marriages in Las Vegas (Firat, 2001/3). And there are the more stark differences that can be traced to the economy, society, and polity of individual cities; differences that determine not just the material standard of living in a city but also who the city believes you can love and hate, as well as whether you have a say in the way it is governed.

Several of these diverse, and often very visible, characteristics of a city are the result of larger processes. In the decades after the communication revolution much attention has been paid to the processes of globalization. The World Cities literature, in its earlier phase at least, tended to be preoccupied with generating hierarchies of cities, seeming to imply that cities could move up or down this hierarchy (Taylor, 1997). Others have been more concerned with the precise processes in Global Cities (Sassen, 2009). In either case the focus has been on the command and control centres of globalization. But the economic power of these centres emerges, in part, from the resources they tap through cities in other parts of the world. The Y2K problem in the programming of computers of the global North at the turn of the century was addressed largely by tapping technical manpower from cities in the global South, often accounting for a major component of the software earning of the countries where these cities were located (Kumar, 2001). Even before the communication revolution, global garment brands with their command and control centres in major Western cities marketed garments manufactured in cities in the developing world, with countries like Bangladesh being major beneficiaries (Rhee, 1990). A natural corollary to the literature on global cities must then be, as I have argued elsewhere, an understanding of what can be termed Resource cities (Pani, 2009).

The differences between the two ends of globalization has been so great as to lead to a well-established, if misleading, binary of global cities versus the megacities of the South. But globalization is not by any means the only set of processes affecting the nature of cities. Cities are often the urban ends of processes that have occurred in different historical situations. In medieval India cities were often large fortresses in which people from the neighbouring rural areas took refuge in times of war. In more recent times, cities, especially in the global South, are often places to which rural people move, temporarily or permanently, in search of work, as can be seen in the rapid process of urbanization in China (Zhang & Song, 2003). At times this movement can be forced on an entire village, as when it is absorbed, willingly or otherwise, into a horizontally expanding metropolis (Zhang & Wang, 2003). Other processes could lead to villages themselves being transformed into cities. This could sometimes happen quite silently when a village grows enough, in population and other ways, to cross the official dividing line between the urban and the rural. And the growth can continue quite dramatically as when Shenzhen in China was rapidly transformed from a small town into a globally recognized metropolis (Yang, 2005). And if we move beyond the economic to the social, political and cultural realms, there are an even larger number of processes that play themselves out in cities.

The influence of territory

Conventional attempts to generate a framework that can capture diversity in the urban have concentrated on placing these urban processes into a territorial straitjacket. Territorial boundaries are drawn around a city in ways that capture the entire contiguous geographical area that is believed to have urban characteristics. The bounded area is then taken to be entirely urban, though it is not unknown for the territory to include rural characteristics, including what has come to be known as urban villages (Liu & He, 2010). The study of the city is then expected to explain all that happens within these boundaries, largely to the exclusion of what happens beyond them.

This approach has a "territorial trap" built into it. John Agnew has outlined this trap in the form of three assumptions that are implicit in the use of territories in political economy: the state has total sovereignty over its entire territorial space; there is a clear distinction between the domestic and the foreign aspects of life, that is, between what happens within the territory and what happens without; and that the economy and society are contained within state boundaries (Agnew, 1994). The extent of this territorial trap is quite evident in national economies. There can be influences within a nation that the state has little control over; in the era of globalization the distinction between domestic and foreign can be blurred in individual lives; and the economy and society of a nation rarely function in isolation.

A territorial trap exists in the realm of cities as well. Local manifestations of the state, which can range from elected local governments to parastatals or even national governments, can, and often do, claim sovereignty over the entire city. They claim the sole right to lay out the laws that the city is governed by, though in reality this need not always be the case. In the south Indian city of Bengaluru (formerly Bangalore) manufacturers produce garments for global brands, like GAP. These brands, being extremely sensitive to consumer sentiments in Western cities, insist the manufacturers follow clear labour and environment standards, such as the SA 8000. They also hire social auditors to ensure these norms are followed within the manufacturing units. With the fear of losing the contracts that ensure they exist, the manufacturers follow these norms very scrupulously. As far as working conditions within the garment manufacturing units in Bengaluru are concerned the sovereignty of the state has been quietly overtaken by the demands of global brands catering to customers in culturally very different cities (Pani & Singh, 2012).

The sharp demarcation that planners, and some others, seek to draw between what exists within a city and outside is also somewhat exaggerated. What happens in a city in one part of the world can be very clearly influenced by events elsewhere. Terrorist acts are more often than not influenced by actors reacting to conditions in another part of the world. 9/11 was hardly provoked by anything specific done by New Yorkers. And the blurring of a city's boundaries is not a matter of such dramatic and dastardly acts alone. In many parts of the global South, urban centres continue to be marked by some influences of the rural. This is most stark when a village is newly absorbed into a horizontally expanding city, but some of these influences can continue in varied forms for decades, and even centuries. Cities in India continue to have a prominent place for the celebration of harvest festivals, such as Pongal in the south Indian metropolis of Chennai.

The assumption that the economy and society of a city can be best understood in isolation is also a very tenuous one. An effective understanding of a specific urban economy or society must necessarily find a place for the events that transform that city, and these events often occur outside that city. The rebellion of French students in 1968 and the Woodstock festival a year later contributed to a sense of fashion in Western cities that celebrated denim and other fabrics and clothes of the Western working class. Recognising that this demand could be met at a much lower cost in cities of the developing world, a large part of the manufacture of the garments sold by global brands shifted to the global South (Gereffi, 2001). This transformation prompted by changing demand patterns in the Western world provided a new direction to the economies of several cities in the South. In addition to the economic effects of this transition, the exposure to the new fashion also had its social influences. The social imagery generated by the new fashion created a desire for these garments. This demand was met by pieces of garments, which had failed to meet quality standards, being sold at throwaway prices. The extent of this influence can be seen in the widespread use of the Western nightgown as a dress worn by working class women in public in several Indian cities. And these patterns of parts of cities being influenced by events and cultures in other parts of the world are not a matter of Western influences in cities of the developing world alone. There can also be social influences in the other direction, such as the predominantly Pakistani settlement of Bradford in England (Pieroni, et al., 2008).

The socially and economically porous nature of the territorial boundaries of cities has its effects on the nature of the city. The bounded areas that constitute the city as a whole typically include a variety of places that have emerged from very different processes. In parts of the global South information technology parks that represent the local end of high technology globalization exist alongside much poorer colonies of workers in lower-end industries (Benjamin, 2010). Both these areas exist cheek-by-jowl with the remnants of villages that have been absorbed into cities; villages that often continue to reflect some of their rural origins. The coexistence of these groups is not always orderly or even peaceful. The formality of one part of the city may not be shared by another part of the city; the economic status of different parts of the city will almost always differ, which, in turn, may or may not be reflected in social status. The ethnic composition of each part of the city could also be very different, particularly when migrants from one part of the world celebrate their original culture in a part of a city in another part of the world. And the responses to this diversity can add to the differences, as when groups with some common economic and social features prefer to enclose themselves into gated communities.

The immediate response to these diverse areas is to treat them as fragments of the larger territory that defines the city. The territory of the city can be seen to consist of smaller geographical areas which can have a great deal of autonomy. It is not unknown for an elite group of persons to live an entire life in one part of a city without visiting a slum in another part of that city. The part of the city that group lives in would have clearly defined economic and social boundaries. These areas can also gain official status when they coincide with a ward or a borough. But these fragmented territories of a large city break each of the assumptions that go to make the territorial trap. The first assumption of this trap, that the state has complete sovereignty over such a local territory, is often even more tenuous in a fragment of a city than it is for the city as a whole. The literature on informality in cities of the global South points to the extent to which everyday practices do not always accept the sovereignty of the state. The second assumption, that a clear distinction can be found between what exists within such a territorial fragment of a city and what happens outside, is also fragile. Take the case of the first technology research centre of the American multinational GE to be located outside the United States. The urban territory defined by the John F. Welch Technology Centre in Bengaluru in south India may appear to meet the condition of being different from what exists outside its boundaries in other parts of the city, and some of the geographically nearby villages. But this difference is the result of the fact that this territory relates much more closely to conditions further away, indeed on the other side of the globe. The experience of working in this space would, in fact, be similar to that of centres of GE in American cities. The third assumption of being able to conceptualize this territorial fragment in isolation is even more difficult to sustain. The very existence of such a space can be determined by decisions taken in the command and control centres of GE. Indeed, if the John Welch centre were to be moved to another city it could well take its urban identity, if not a large number of persons working there, to its new location.

Such continuous change in the urban ensures that the form at any given point of time has an element of temporariness to it. The dynamic nature of urban processes in an interconnected global space can also see small settlements, such as those around factories in the global South producing for global brands, having the features, particularly architectural, that are usually seen in large metropolises. This limits the relevance of the traditional classification of town, city, or metropolis. The view of the city as a completely urban bounded space too comes up against the continuously changing geographical and other spaces of the urban. There is clearly much to be said for the view that the urban cannot be seen as an urban form, settlement type or bounded unit (Brenner, 2013).

The urban experience thus does not fit neatly into territorial boundaries. Instead the relevant boundaries keep changing according to the activity being considered. Economic considerations could see boundaries that protect the link between a fragment of a city in a less developed country with a fragment of a city in a geographically-distant developed economy. Cultural considerations, such as the celebration of a festival, could see the boundaries of the same geographic fragment in a developing country being extended to geographically neighbouring areas. Elections could see political boundaries gaining greater importance, just as social upheavals can generate new boundaries of their own. The relevant boundaries are thus prone to change depending upon the issues and actions that are being considered. Add to this the fact that the boundaries within each domain are themselves prone to change. The expansion of globalization in a city in the global South would change the economic boundaries within that city, with its possible effects on social and other boundaries. The emergence of a new locally developed economic engine of growth too would have its effects on boundaries, as could social changes such as ethnic conflicts. The boundaries of the urban, and what happens within them, are thus continuously changing.

The easy way out of the challenges posed by the territorial trap is to do away with territory altogether. This would be possible if we accept Lefebvre's hypothesis that society has been completely urbanised (Lefebvre, 2003). This contention has gathered momentum in the decades after it was first made in 1970. The information and communication technology revolution has contributed to the view of Brenner and Schmid that "Today, urbanization is a process that affects the whole territory of the world and not only isolated parts of it. The urban represents an increasingly worldwide, if unevenly woven, fabric in which the sociocultural and political-economic relations of capitalism are enmeshed" (Brenner & Schmid, 2014, p. 752). It would appear that no part of the world is completely untouched by the influences of the urban, whether it is through capital, knowledge, fashion, or any other instrument. Even wildlife in the most remote parts of the planet can be brought into a conceptualization of the world developed in the urban mind through the creation of wildlife sanctuaries. The indigenous communities living in these forests often have little influence on norms that are laid out for them to follow, typically by urban environmentalists and policy makers. Such an all-encompassing concept necessarily has differences built into it; indeed Brenner and Schmid argue that planetary urbanization increases this differentiation.

Using a single concept of planetary urbanization to explain everything, though, runs the risk of not being very useful in understanding specific processes around the world. The concept may have its uses in recognizing urban processes that extend far beyond the territories of cities, but it can be misleading when it implies that all processes in the world are a part of capitalist urbanization. This would suggest overreach at least as far as some of the more remote communities in the world are concerned. There remain communities based on agriculture, or even hunting and gathering, with very few traces of the urban in them. There are communities in the Himalayas that have little space for the urban. Away from remote areas too there are settlements that would not fit conditions of capitalist urbanization. It is difficult to consider agricultural villages in south Asia, which still have a prominent place for extra-economic coercion right down to honour killings of daughters who marry outside traditionally defined norms, as a part of urban capitalism. Even if we were to modify our definitions to somehow classify these villages as a part of planetary urbanization, there is little doubt that the experience of these predominantly rural settlements will be very different from that of Mumbai, let alone New York. In bringing all the possible variations − including the rural − under its umbrella, the concept of planetary urbanization does away with some of the specific tools needed to understand individual urban experiences.

The discourse on urban territories emphasises the difficulty in defining the boundaries of the urban. At one extreme we can demarcate geographical boundaries of the urban that ignore the non-urban within them, while at the other extreme we can extend the boundaries to the entire planet thereby once again diluting our ability to distinguish between the urban and the non-urban. The difficulty urban theory faces in dealing with territory is thus closely associated with a second challenge; that of coming to terms with diversity. As the vast body of empirical work grows even larger, it throws up numerous specific insights, sometimes leading to what is seen as a further difficulty in coming to terms with the diversity of the urban experience: the difference between the general and the particular.

Responding to difference

Efforts to find a general theory that would explain diverse particular experiences have taken different routes. The most prominent route has, arguably, been to define a city by its dominant characteristic. This has contributed to the phenomenon of labelling cities as global cities, fragmented cities, neo-liberal cities, and several other terms. Arriving at an overwhelmingly dominant characteristic has, however, proved to be quite difficult. No matter how important a feature of a city may appear to be, it necessarily suppresses several of its other aspects. The concept of a global city has undoubtedly helped us understand the urban in the process of globalization (Sassen, 2009), but it undermines activities in parts of the city that are relatively untouched by globalization. There may also not be unanimity on the choice of the dominant feature of a city. Some might see twenty-first century Rome as a historical city, while others would emphasize its prominent political role in Italy and even Europe. Some might see Varanasi as a major site of traditional Hinduism, while others would celebrate the Muslim musical traditions that reside there (Lee, 2000). And this is even without getting into the more openly contested cities like Jerusalem. Differences in the perception of the dominant feature of a city can thus emerge not just from competing empirical realities but also from individual perceptions of what is the most important aspect of a city.

One route to bypassing the challenges that arise when seeking a dominant feature is to focus on inherently comprehensive characteristics, like the size of a city. The concept of megacities, by focusing primarily on size, has space for all the diverse characteristics that fit into the large cities of the global South. Mumbai houses both large slums and India's financial capital; it offers most of its children no more than the streets as playgrounds, even as it has become a major centre of world cricket (Bateman & Binns, 2014). This diversity encourages a host of processes, not all of which are formally regulated. The megacity thus includes considerable scope for informality, which has been analysed in substantial empirical detail. This work has also generated concepts like occupancy urbanism (Benjamin, 2008). But the journey from these concepts to a larger theory has been halting, and not without good reason. The informal is, strictly speaking, a negative criterion that refers to all that is not formally regulated. It is thus an umbrella term that covers a wide range of processes and features. It includes the illegal as well as unregulated activities that may be entirely legal; it includes products built from crude technologies as well as the repair of the most high-technology mobile phones; it encompasses poor squatters as well as those owning expensive, improperly acquired real estate. The concept of megacities allows us to capture this extensive empirical detail, and thus helps us understand the "how" of these cities, but it is not always as successful in explaining the "why" of what happens there. What megacities gain from the comprehensiveness provided by making size their primary feature, they lose in terms of the analytical ability to explain the processes that make a city.

Faced with the difficulty of finding a single concept that would explain all that happens within the territory of a city, there has been a preference to simply celebrate the diversity. The idea of fragmented cities matches the empirical reality of cities consisting of multiple, often very different, places. But moving from this empirical reality to how the different fragments interact with each other provides its own challenges. The specific fragments would vary from city to city. The role the Ram Lila celebrations play in Delhi would not be shared by Berlin. Even when the same fragments exist in multiple cities, their specific urban roles could be very different. Financial networks in cities with a prominent place for the informal would not necessarily be similar to the mechanisms of the financial centres in the global North. Indeed, given the number of factors involved, and the varying degree of significance of each of them, as well as the prospects of very different relations between them, there is a strong possibility of each city being quite unique. This has contributed to the idea that all cities are unique, and there is nothing unusual about this uniqueness.

Jennifer Robinson celebrates this uniqueness, even as she recognises it to be a part of a larger process, through her concept of ordinary cities. She understands ordinary cities as unique assemblages of wider processes – they are all distinctive, in a category of one. Of course there are differences amongst cities, but ... these are best thought of as distributed promiscuously across cities, rather than neatly allocated according to pregiven categories ... ordinary cities bring together a vast array of networks and circulations of varying spatial reach and assemble many different kinds of social, economic and political processes. Ordinary cities are diverse, complex and internally differentiated.

(Robinson, 2006, p. 109)

In highlighting the uniqueness of the particular in the midst of general processes, Robinson recognises the need to take the debate beyond the traditional divide between the empirical particular and the theoretical general. Her work does provide a sense that all urbanists can learn from each other. The vast body of empirical studies throws up varied insights, with those from one urban experience possibly proving useful in exploring another such experience. But Robinson stops short of addressing the methodological issue of how exactly a specific case of the urban fits into larger urban processes. While her argument for understanding the empirical detail of each case of urbanization is well founded, she pays very little attention to how the different experiences of the urban fit in with each other. She would appear to be, perhaps unintentionally, supporting the case for an unending series of studies of individual urban situations at the cost of adequate attention being paid to the larger lessons that emerge from the studies taken together. Thus, in practice she is undermining the need to see the empirical particular in the context of larger urban processes. Indeed, the idea of ordinary cities has sometimes been presented as one end of the spectrum with the other end being a general theory of the urban. As in the case of the territorial trap the challenge of dealing with differences has also tended to fall into two extremes. One end has tended to be preoccupied with the particular case, while the other has been in search of general categories based on the perceived dominant characteristic of a city.

Taking processes on board

One reason for the sharp, and not always helpful, binaries in responses to the territorial trap as well as the challenge of differences, is universalism in explanations of the urban. There is an implicit assumption in much of urban studies that a theory must explain all situations across time and place. This assumption demands both comprehensiveness and consistency. The search for comprehensiveness typically takes the form of first drawing territorial boundaries with an implicit expectation that all that exists within those boundaries will be explained. But, as we have seen, the comprehensive explanation of specific urban phenomena can take us well beyond these boundaries, leaving us with an apparent choice between either conceptually ignoring what happens outside the boundaries or doing away with boundaries altogether. Similarly, the search for consistency has been implicitly assumed to be a need for different situations in different urban places to all fit into a single theoretical explanation, usually based on a dominant feature of the urban situation, such as the effects of globalization. When the theory fails to do so, particularly when the levels of empirical detail are quite high, there has been a tendency to treat each case as a separate conceptualization leading to the celebration of the uniqueness of each city. The larger picture is then usually confined to the network of connections between cities rather more than what happens in them. The response to the need for consistency can then take the form of either focusing on a dominant feature and underplaying, if not ignoring the rest, or falling back on concepts that celebrate uniqueness like that of ordinary cities.

A changing urban situation can increase the challenges of both finding the appropriate boundaries as well as responding to the differences within urban spaces. The effective boundaries of a command and control centre can keep changing as business opportunities change. A global garment brand based in a city in the global North might find it profitable to tap a manufacturing source in a city in Bangladesh one month and shift it to a city in Vietnam some months later. Such shifts may well increase the differences in the activities within those boundaries. Despite efforts of global brands to impose labour and environment standards, there are a number of cultural elements, including the gender sensitivity of local women workers, which would differ from country to country. The inevitability of such change makes as strong a case as any to follow Lefebvre in treating the urban as a process. The difficulty in trying to capture this essentially dynamic process through theories that are static over time and place is quite evident.

At the heart of this difficulty is the fact that, by definition, a static theoretical system has little time for dynamic change. A static theory would, by its very nature, tend to underestimate change even when it does not ignore it altogether. There is a tendency to see change largely in a comparative static framework. In this framework change can be monitored by studying the same urban phenomena across different places or over time. As these studies mature in their empirical detail, though, the differences typically grow beyond what can be convincingly captured by comparing static models of two or more places at the same time, or the same place at two or more different points of time. The remarkable study of London, New York and Tokyo by Saskia Sassen does contribute significantly to our understanding of the effects of globalization on the urban, yet it cannot claim to explain the *entire* process of dynamic change in these three globally dominant metropolises (Sassen, 2009). Rather than using essentially static frameworks to understand a rapidly changing urban reality, we would be better served by dynamic concepts of the urban process.

A significant step in this direction has been taken by Scott and Storper when they build an understanding of the urban around the process of agglomeration. Observing that agglomeration has always been a central feature of urbanization, they go on to argue that it "is the basic glue that holds the city together as a complex congeries of human activities, and that underlies—via the endemic common pool resources and social conflicts of urban areas—a highly distinctive form of politics" (Scott & Storper, 2015, pp. 6-7). Each process of agglomeration is distinct from other processes in the city as well as from other processes of the same urban space, though there is scope for some overlap. The coexistence of multiple processes in a larger urban space, each with its own distinctive features, ensures the process of agglomeration is accompanied by polarization. These processes in turn influence the interactions and the relations with land. For Scott and Storper urbanization revolves "around processes of agglomeration cum polarization and associated interactions within the urban land nexus" (Scott & Storper, 2015, p. 9).

This definition of the urban provides two important tools to help us understand the urban in the global South: the process of agglomeration cum polarization, and the difference between the urban and the city. The use of the phrase "agglomeration cum polarization" may suggest that polarization is no more than the flip side of agglomeration, and vice versa. There are no doubt long periods when this is in fact the case. The differences between different processes of agglomeration does generate increased polarization. This is often visible within workplaces in the global South when workers migrating to the city from different regions tend gravitate towards each other. Yet there are also times when the two processes work against each other. There are situations when the agglomeration of different ethnic groups to the same workplace can reduce the social polarization between the two groups. What usually causes greater urban turbulence is when extreme polarization generates urban tensions on a scale that cause a mass exodus of an "outsider" ethnic group. In those times, polarization can weaken the processes of agglomeration. It is thus important to emphasize that while agglomerations do lead to polarization, not all polarization is a result of agglomeration. Indeed, there are times when some forms of polarization can result in the weakening of agglomeration. In view of this element of autonomy to the processes of agglomeration and polarization, it may be better to refer them as agglomeration *and* polarization, rather than agglomeration *cum* polarization.

Another consequence of this approach to definition of the urban process is that it allows for a distinction to be made between what is found in a city and what is urban. Scott and Storper recognize that there can be aspects of a city that are not a part of urban processes. They point out that while there are usually many poor people in cities it does not mean that all aspects of poverty are urban in nature. This approach can also be interpreted as a way out of the territorial trap. Rather than operating with one fixed boundary of the urban we can postulate separate boundaries for each urban process. The boundaries of individual urban processes can not only extend beyond the administrative boundaries of the city, but could also exclude parts of the administrative territory identified with a city. This interpretation breaks out of the three assumptions of the territorial trap. It does not require that the state has total sovereignty over territorial space. In our example of a garment manufacturing unit in a city in the global South, the labour standards prescribed by the global garment brands are enforced in a way that bypasses the local or national state. There is also no clear distinction, certainly not in a territorial sense, between the domestic and foreign aspects of life. Functioning during working hours within the boundaries laid out by the global garment brand can blur the distinction between domestic and foreign, with the domestic manufacturer often enforcing norms of foreign origin. And any assumption that the economy and society are confined within boundaries laid out by the state runs contradictory to the everyday practice in the garment manufacturing unit.

In taking their conceptualization – of the urban as a processes of agglomeration cum polarization – to the challenge of differences within urban spaces, though, Scott and Storper are somewhat less successful. Their route to reducing the complexity of the extent and variety of differences in urban reality is to shift the focus to the sources of these differences. They indicate that the entire gamut of differences across urban situations can be captured by the major sources of these differences and the interactions between these sources. Built into their argument is the belief that the diversity in urbanization can be traced to five major sources: overall levels of economic development, rules that govern resource allocation, structures of social stratification, cultural norms and traditions, and overarching conditions of political authority. It is possible, though, that the extent to which these categories actually reduce the differences to be considered may be somewhat exaggerated. Each of these categories contains a vast range of specific sources of variation in the overall urban situation. An apparently technical norm like the rules of resource allocation can range from clearly defined rules based on consistent principles to the anarchy and lawlessness of Al Capone's Chicago. And when the rules that can influence urban resource allocation are anarchic they are not of much use in explaining the differences in resource use in cities. Some researchers have developed umbrella terms that can capture the entire range of such anarchic rules in some types of cities, the most notable among them being informality. But such terms are so broad, and consist of so many different types of urban functioning, that they are not always very useful in gaining insights into the specific nature of urban informality in a city. What is more, it is not entirely clear that all the sources of differences in the urban can be captured in this five-fold categorization. Broad as this categorization is, it is not impossible to identify other elements that influence a city. It could be argued that the history of a city contributes to its particular characteristics. There are vast differences between Sassen's global cities of London, Tokyo and New York; differences that have in part been caused by their history. And it is not unknown for historical events to transform a city altogether. The making and the breaking of the Berlin Wall contributed to some of the uniqueness of that city, and the bombing of Hiroshima and Nagasaki transformed those two previously peaceful urban places forever. This shift from differences to the sources of differences thus does not necessarily diminish the difficulty in addressing diversity.

The benefits to addressing the territorial trap, provided by the recognition of the urban as a process, does not, in the Scott and Storper formulation, extend to an explanation of the differences across cities. At least a part of the blame for this deficiency can be traced to the fact that they do not fully explore the implications of the distinction between the urban and the city. As pointed out earlier, they do recognize that not all aspects of a city can be attributed to the process of urbanization. But the distinction between the city and the urban can go further. There may even be elements of the rural in the city. This is particularly true of cities in the global South that have typically grown horizontally to envelope villages. While villages that are now within the city do change, it is not as if they lose their rurality overnight. This challenge is addressed later in the book by distinguishing between the urban as process and the city as place. The rest of this introductory chapter is confined to the methodological implications of the shift from the static conception of the urban to that of the urban as a process.

Urban process and action

The effort to develop methods to capture the urban process could benefit from a glance at the way philosophers understand process and its implications for methodology. The idea of the urban being dynamic and continuously changing is consistent with the premise of process philosophers "that the dynamic nature of being should be the primary focus of any comprehensive philosophical account of reality and our place within it" (Seibt, 2016). When reality is seen as a dynamic process any picture we develop of it is necessarily a snapshot of the process at a particular point of time. For this snapshot to help us understand the process it would need to contain all the elements that could influence that process. This would include not just the dominant, and hence easily identifiable, elements of the picture but also seemingly less significant elements that can either grow in influence or simply act as a trigger for more substantial urban transformation. The picture must then capture not just the elements of change at that point of time but also potential sources of change at a later date. This approach militates against seeking to understand change in terms of the major features of the city at a point of time or through predetermined sources of change. Even when the sources of change in a particular situation are well known, the nature of the transformation could itself vary. A snapshot that is of relevance to understanding change in the process must then be finely grained. As every potential element of change must find a place in it, it would have to identify the minutest detail. This in turn demands that our unit of analysis is small enough to capture the slightest change in the process.

Our analysis of the urban process must then move through four conceptually distinct methodological stages. First, it must clearly distinguish this process from others that may be occurring in the same location at the same time. Second, it must choose a unit of analysis that would be capable of capturing all the details of change in that process. Third, it must use these units to develop a detailed snapshot of where the process is at that time and place. And finally it must use this situational analysis to understand elements of the larger urban process, including the role of space and place.

In the first stage of defining the urban process as distinct from other processes, we would do well to begin with the conceptualization of Scott and Storper, and go on to expand it. They see cities as consisting of two main processes:

agglomeration-cum-polarization, and that of the unfolding of an associated nexus of land uses, locations and human interactions. Through history the process of agglomeration has been the mark of all cities. This is quite evident when we recognise that agglomeration need not be prompted by economic conditions alone. Agglomeration prompted by military necessities has in the past led to fortress cities, just as religious considerations have led to the creation of other cities. There is thus much to be gained from Duranton and Puga, who see the micro foundations of urban agglomeration economies in sharing, matching and learning (Duranton & Puga, 2003). Agglomeration aids the process of sharing indivisible elements. It also enables individual specialization which allows the benefits of the resultant increased productivity to be shared. The matching of economic elements, such as workers and jobs, is also enabled by agglomeration in two ways: the quality of each match is expected to improve with the increase in the number of agents involved in the matching; and stronger competition generated by agglomeration leads to a more efficient utilization of fixed assets. Agglomeration also helps learning by bringing together a large number of people and their ideas. This conceptualization of agglomeration can be extended to cover a more explicit recognition of social processes, such as the matching in urban agglomerations that also works in the realm of social groups. Sharing, matching and learning bring with them the questions who do we share with, who are involved in precise matches, and who do we learn from or teach? The answers to each of these questions have an element of polarization built into them.

The need for a broader conceptualization is greater in the process accompanying agglomeration-cum-polarization, that is, in the perception of Scott and Storper, the unfolding of the nexus between land uses, locations and human interactions. While these may be some of the important consequences of agglomeration and polarization, they are not the only ones. These consequences go beyond the land nexus to also cover a variety of other spaces, including those that exist entirely in the imaginations of the urban. *The urban can then be distinguished as the processes of agglomeration and polarization and the entire set of interactions they generate between people, as well as their consequences in multiple spaces.* This formulation may suggest the need for a broad all-encompassing concept, like planetary urbanization, to cover all the possible consequences of agglomeration and polarization. But, as has been noted earlier, the fact that the consequences of agglomeration and polarization can potentially exist at any point on the planet does not mean that the entire planet must necessarily be considered urban. Even when urbanization at different points of time influences different parts of the world, we could still distinguish between what is urban, and what is not, in a particular situation.

In treating agglomeration and polarization and their consequences as the urban process we must explicitly recognize that this does not tell us everything about a city. Each city also has a past, and a memory of that past. This memory can sometimes be overwhelmed by a single aspect of that past. Postcolonial interpretations of cities, at times, focus so much on their colonial past that they tend to ignore some of the non-colonial elements in that history. At other times, the memories of a city are rather more diverse. There can be memories of a city that look back nostalgically at what a city once was, or is remembered to have been. Thus, while the processes of agglomeration and polarization, and their consequences can be said to capture the urban process, they are not all that a city is. Conversely, when we take into account the shifting boundaries of urban processes, a city is also not all that an urban process is.

In the second methodological stage, the unit of analysis needed to capture each urban situation would need to satisfy two criteria: it must relate to the processes of agglomeration and polarization as well as capture the minutest detail that could potentially trigger changes in other dimensions of the urban process. Social scientists tend to view the economy and society as a collection of individuals or of groups. The units of analysis are then essentially the individual or a group, such as caste, race or class. In this perception the individual would be the smallest unit of analysis. When reality is seen as a process, though, this need not be true. The individual is herself continuously changing; she is always "growing". This complex process is "composed of various physical interactions, experiences, feelings, moods, and actions in their systemic interrelationship" (Seibt, 2016). The Wimoa who left her village to join her husband in a city of the global South, is not the Wimoa who has found her feet in the city a couple of years later. The unit of analysis must then be capable of capturing the change that is taking place in the individual as well.

In the context of the requirements of urban processes, actions would be an effective unit of analysis. The three micro foundations of agglomeration – sharing, matching, and learning – all involve specific actions. The consequences of these actions, in turn, involve other actions, including those required to deal with specific urban situations. The actions themselves could be on differing scales. An agricultural worker in a remote part of India may explore the possibility of finding a job in a city, in order to debate with himself the virtue of joining the process of agglomeration. If he decides to actually find a job, and take it, the scale of the action would be larger though still in the realm of the individual. If he were to then identify himself with a social group that would help him in his existence in a city he could find himself carrying out tasks for that group, and the scale in which that action operates would be even larger as he becomes a part of the group and its action.

Treating actions as the unit of analysis impacts the third methodological stage of developing a meaningful snapshot of the process of agglomeration. Actions can be conceptualized at multiple levels of aggregation, from the individual to larger social groups. The use of this flexibility can determine the quality of the analysis as groups are not just an aggregation of individuals. Groups might bring individuals together but they can generate behaviour of their own. Mobs have been known to behave in ways that each of the individuals who constitute them may not have behaved on their own. Individuals are also known to be influenced by the class, or any other social group, they come from. Each level of aggregation of individuals in the process of agglomeration cannot then necessarily be derived from another. Just as a group is not merely an aggregation of individuals, individuals cannot be treated as stereotypes of the groups they belong to. Much attention must then be paid to the relevant social level of the action that is being analysed. This would normally be determined by the question that is being asked. The action of an errant driver would lend itself to an analysis focused on the individual, while that of a mass protest would be better explained by focusing on groups. This combination of individual and group actions form a part of the larger picture of the urban as a set of interrelated actions that constitute the processes of agglomeration and polarization and their varied consequences across multiple spaces.

The question that is asked of a specific situation in the process of urbanization would also influence the rest of the snapshot that is generated. When the question relates to one particular aspect of the urban process a set of actions would need to be analysed, while another question would require attention to be paid to another set of actions. If our concern is with racial discrimination generated by the urban process, our focus would tend to be primarily on the actions influencing race relations and only secondarily, if at all, on the actions involved in, say, repairing the city's airport. If our interest was instead on the role of air travel in the process of agglomeration, the actions involving the repairing of the city's airport would gain primary significance with only a secondary place for unrelated actions affecting race relations. The questions we seek to answer determine which part of a finely grained picture of actions we focus on. This approach thus escapes the territorial trap as it does not need to make any assumptions about the sovereignty of the state over any particular territory; it does not require any distinction to be necessarily made between the domestic and the foreign; and it certainly does not see the economy and society resulting from agglomeration as being confined to any predetermined territory.

The central role for the questions we ask about a particular urban situation determines the fourth methodological stage of our analysis, that is, how we relate individual snapshots to the urban process. The questions may be designed to generate an empirical picture: the how, where and when of a particular urban situation. Alternatively they may seek to analyze a particular situation, exploring the why of the urban process. The choice of the question would reflect the perspective of the researcher, but this choice need not be completely anarchic. It will also be influenced by the intellectual milieu from which it emerges. The questions would emerge from particular interpretations of the urban process, and their answers would ideally lead to the strengthening, modification or rejection of those interpretations. This is not to suggest that this is entirely an exercise of the intellectual community. A migrant worker from a village who comes to a city in search of work has a particular perception of the urban process which helps her interpret specific actions of others in the agglomeration process as opportunities. Her experience when responding to those opportunities will determine whether her interpretation of urban actions is strengthened, modified or rejected. The snapshots of the urban situation, more specifically the parts of it that we choose to focus on, thus play a critical role in determining how we see the urban process.

Relating snapshots of specific situations to larger urban processes has its implications for the way we see the relationship between the particular and the general. The focus on specific situations is designed to capture their uniqueness, without claims to any individual situation representing the urban process in other times and places. The unique character of each situation does not, however, imply that there is no element of a particular urban situation that can emerge in another urban situation; that the relations seen in one urban situation are entirely irrelevant in another. While there is certainly reason to be sceptical of approaches that mechanically transfer all the lessons from one situation on to other times and places, there is a case for a more differentiated method of seeing which lessons from a particular situation can be transferred to another and which cannot. There could be specific relationships between a small set of factors that hold across situations, there could be other relationships that are unique, and there could be still other relationships that hold for some cases and not for others. These diverse influences of one situation on another can be clarified by distinguishing between models, arguments, and methods.

We can take a *model* to refer to an abstraction of a particular relationship between different actions and their consequences that is expected to work across time and place. The relationship the model captures may be distinctly noticeable, such as the one between the pace of migration and the size of the city. The abstraction would ideally be a universal principle that will work in all situations, such as the Proximity Principle that will be defined in the next chapter. Yet it is also possible to define a model in a way that it will hold only for specific types of urban situations. It is theoretically possible that some specific situations may be explained in terms of a single model, but it is far more likely that several models exist simultaneously, each explaining but one part of a specific situation in urban reality.

We can treat a more complete explanation of a specific situation as an *argument*. In the rare case where a single model explains an entire urban situation, or the situation of a city, there would be no difference between the model and the argument of that situation. In the far more likely scenario where this is not the case, an argument would consist of more than one model. In bringing models together to explain a particular situation there will be a prominent role for what are considered to be objective facts. Statistical exercises may be used to capture specific empirical relationships. But the argument would have to enter the domain of judgement as well. The very choice of models would involve an element of judgement, as would the precise ways in which different models are put together. It is important to stress that since the purpose of this argument is to explain, the use of judgement would ideally be restricted to what the researcher believes to be true; it would not extend to what the researcher would like to be true, howsoever desirable that may be.

We can take method to refer to the entire set of rules to be followed in understanding the urban and the city. It includes both the means of deciding the questions that are to be raised as well as the process of answering them. It involves bringing together models and arguments to understand a particular urban situation or a situation in a city. It could go beyond explaining existing situations to predicting what is likely to happen. It can go even further to what the researcher would like to happen and what it would take for that to be realised. This exercise would use both facts that are widely believed to be objective, as well as judgements. Its approach to subjective judgements would not necessarily be to reduce their relevance or to convert them into objective criteria. It would, instead, recognise their value and seek to improve the quality of subjectivity. While judging the quality of subjectivity would require the benefit of hindsight, it is possible for individuals who have made successful judgements to develop a reputation for being able to do so. They would then find it easier to gain endorsement for their proposed actions, and influence the course of urban processes as well as those of the city.

References

Agnew, J., 1994. The Territorial Trap: The Geographical Assumptions of International Relations Theory. *Review of International Political Economy*, 1(1), pp. 53–80.

Bateman, J. & Binns, T., 2014. More than Just a Game?: Grass Roots Cricket and Development in Mumbai, India. *Progress in Development Studies*, 14(2), p. 147–161.

Benjamin, S., 2008. Occupancy urbanism: Radicalizing Politics And Economy Beyond Policy and Programs. *International Journal of Urban and Regional Research*, September, 32(3), p. 719–729.

Benjamin, S., 2010. The Aesthetics of "the Ground up" City: Some Insights from Bangalore. *Seminar*, 612, pp. 33–38.

Brenner, N., 2013. Theses on Urbanization. *Public Culture*, 25(1) Duke University Press, pp. 85–114.

Brenner, N. & Schmid, C., 2014. The "Urban Age" in Question. *International Journal of Urban and Regional Research*, May, 38(3), pp. 731–755.

Cohen, B., 2004. Urban Growth in Developing Countries: A Review of Current Trends and a Caution Regarding Existing Forecasts. *World Development*, January, 32 (1), p. 23–51.

Cohen, B., 2006. Urbanization in Developing Countries: Current Trends, Future Projections, and Key Challenges for Sustainability. *Technology in Society*, January April, 28(1–2), p. 63–80.

Duranton, G. & Puga, D., 2003. Micro-Foundations of Urban Agglomeration Economies. NBER Working Paper Series, August, pp. 1–59.

Firat, A. F., 2001/3. The Meanings and Messages of Las Vegas: The Present of our Future. *M@n@gement*, 4, pp. 101–120.

Fisher, W. F., 1997. Doing Good? The Politics and Antipolitics of NGO Practices. *Annual Review of Anthropology*, 26, pp. 439–464.

Gereffi, G., 2001. Global Sourcing in the U.S. Apparel Industry. *Journal of Textile and Apparel, Technology and Management*, 2(1), pp. 1–5.

Gesler, W. M. & Pierce, M., 2000. Hindu Varanasi. *Geographical Review*, April, 90(2), pp. 222–237.

Gleeson, B., 2012. The Urban Age: Paradox and Prospect. *Urban Studies*, April, 49(5), pp. 931–943.

Haque, J., 2022. Caste, Power and Aspiration in Structural Dualism. In: N. Pani, ed. *Dynamics of difference: Inequality and transformation in rural India*. New Delhi: Routledge.

Huang, L., Yan, L. & Wu, J., 2016. Assessing Urban Sustainability of Chinese Megacities: 35 Years after the Economic Reform and Open-Door Policy. *Landscape and Urban Planning*, 145, p. 57–70.

Kumar, N., 2001. Indian Software Industry Development: International and National perspective. *Economic and Political Weekly*, 10–16 November, 36(45), pp. 4278–4290.

Lee, C., 2000. "Hit It With a Stick and It Won't Die": Urdu Language, Muslim Identity and Poetry in Varanasi, India. *The Annual of Urdu Studies*, 15, pp. 337–351.

Lefebvre, H., 2003. *The Urban Revolution*. Minneapolis: University of Minnesota Press.

Liu, Y. & He, S., 2010. Urban Villages under China's Rapid Urbanization: Unregulated Assets. *Habitat International*, 34, p. 135–144.

Mngutyo, I. D. & Jonathan, O. A., 2015. Optimizing the Concept of Place-Making as a Panacea for Informaility in Urban Areas of Developing Countries. *International Journal of Energy and Environmental Research*, December, 3(3), pp. 25–30.

Pani, N., 2009. Resource Cities across Phases of Globalization: Evidence from Bangalore. *Habitat International*, 33, p. 114–119.

Pani, N. & Singh, N., 2012. *Women at the Threshold of Globalization*. Delhi: Routledge India.

Pieroni, A., Sheikh, Q.-Z., Ali, W. & Torry, B., 2008. Traditional Medicines Used by Pakistani Migrants from Mirpur living in Bradford, Northern England. *Complementary Therapies in Medicine*, 16, p. 81–86.

Rhee, Y. W., 1990. The Catalyst Model of Development: Lessons From Bangladesh's Success with Garment Exports. *World Development*, February, 18(2), pp. 333–346.

Robinson, J., 2006. *Ordinary Cities: Between Modernity and Development*. Abingdon: Routledge.

Roy, A., 2011. Conclusion: Postcolonial Urbanism: Speed, Hysteria, Mass Dreams. In: *Worlding Cities: Asian Experiments and the Art of Being Global*. Chichester, UK: Wiley-Blackwell, pp. 307–335.

Roy, A. & Ong, A. (eds.), 2011. *Worlding Cities: Asian Experiments and the Art of Being Global*. Chichester, UK: Wiley-Blackwell.

Sassen, S., 2009. Global Cities and Survival Circuits. In: *American Studies: An Anthology*. Chichester, UK: Wiley-Blackwell, pp. 185–193.

Scott, A. J. & Storper, M., 2015. The Nature of Cities: The Scope and Limits of Urban Theory. *International Journal of Urban and Regional Research*, 39(1), pp. 1–15.

Seibt, J., 2016. *Process Philosophy*. [Online] Available at: http://plato.stanford.edu/archives/fall2016/entries/process-philosophy/ [Accessed 15 October2016].

Taylor, P. J., 1997. Hierarchical Tendencies amongst World Cities: a Global Research Proposal. *Cities*, December, 14(6), pp. 323–332.

Taylor, P. J. & Csomós, G., 2012. Cities as Control and Command Centres: Analysis and Interpretation. *Cities*, 29(6), pp. 408–411.

Yang, C., 2005. An Emerging Cross-Boundary Metropolis in China: Hong Kong and Shenzhen under "Two Systems". *International Development Planning Review*, 27(2), pp. 195–225.

Zhang, C. & Wang, F.-r., 2003. The Countermeasures about Urbanization of Villages in Cities. *Journal of Xi'an University of Architecture & Technology*, 3, pp. 15–18.

Zhang, K. H. & Song, S., 2003. Rural-Urban Migration and Urbanization in China: Evidence from Time-Series and Cross-Section Analyses. *China Economic Review*, 14 (4), pp. 386–400.

2 The making of urban action

A meaningful exploration of Wimoa's actions as she enters the city would require us to first make explicit what we mean by action. In the analysis of social processes actions have been seen from multiple perspectives. From a rights-based perspective the concern has been primarily with the capacity to act, or agency. Critical as this may be to the course of the actions that go into the making of an urban process, it does not quite tell us what an action is. This is a question that has been of greater interest to philosophers who have debated the distinction between action as something that *happens* and action as something that is *done*. This debate has tended to veer towards the latter view with a sense that action necessarily involves intention. There is considerable support in philosophical discourse for Donald Davidson's influential 1980 essay that insisted that actions were in some sense intentional (Davidson, 1980). And there is little doubt about the role of intention in several of the actions in the processes of agglomeration and polarization and their consequences. When exploring the actions of Wimoa, though, our concern would have to be both with something that is done as well as something that happens. Take a possible sequence of actions. She buys a ticket on a bus with the intention of going to the city. The bus happens to break down on the way to the city. During the long wait for the bus to be repaired she gets to know a couple of other passengers who, in turn, know someone who can get her a job in the city. They give her the name of the factory and that of the person she should contact. By the time she reaches the city she is so tired that she only remembers the name of the factory but not that of the person she should contact. In her journey she has intentionally got on a bus to the city; she happened to meet others who could find a place for her in their social network; they intentionally gave her a specific contact in a factory; a contact she happened to forget. From Wimoa's perspective what matters is not just the intentional actions contributing to agglomeration and polarization but also all that happens, as well as the consequence of those actions. In most situations the intentional actions of agglomeration, what happens, and both their consequences are invariably closely interrelated. An exploration of the larger process of urbanization must then cover all that happens in the processes of agglomeration and polarization (whether intended or unintended), their consequences, further actions they generate, and so on.

DOI: 10.4324/9781003196792-2

Extensive as this conceptualization of action is, it may not be broad enough. There are situations in the process of urbanization that would require us to go beyond material actions to the realm of thought. As Wimoa travels in the bus to a city she has never been to, she would be surrounded by men she has never seen before. If she is not used to travelling alone, and has been given the usual advice about strange men, there would be an element of fear of at least some of them. This fear could be accentuated if the strangers belong to a different identity group, say an ethnic group who others in her village have hated. That fear may turn out to be unfounded, and no untoward incident may happen during the journey. Yet the experience of the sustained fear of a particular person belonging to a particular group for several hours in the closed environment of a bus could well contribute to the way she sees that group in the city. In such a case the fearful thought itself has consequences. The potential consequences of thought make a case for treating thoughts themselves as actions. For Wimoa these thoughts would include memories of the fear she felt along the way. This fear would have been influenced not just by what happened during the processes of agglomeration and polarization but also her earlier thinking. Recognition of such elements in the urban process would lead us to the ideas of those, like the Indian political thinker MK Gandhi, who believed that a meaningful conceptualization of action would include thought.

> Karma means any action, any bodily activity or motion. In the [Bhagavad] Gita's definition of the word, however, Karma includes even thought. Any motion, any sound, even breathing, are forms of karma. Some of them we cannot avoid performing. Some of them we perform as a matter of necessity. Some others are involuntary.
>
> (Gandhi, 1969)

The conceptualization of the urban that then emerges is one of a set of interrelated actions – seen as all that happens in the physical and other domains, both intentional and otherwise – that form the interactive processes of agglomeration, polarization, and their consequences.

This treatment of urbanization as a set of interrelated actions is consistent with the arguments laid out by the philosopher Michael Thompson (2010). Thompson sees actions as being explained by other actions. Wimoa's action of getting on a bus is explained by, among other things, the action of her husband going to the city earlier. And it is because she got on the bus that she met the others who drew her into their social network. We can interpret this process as a series of actions and their consequences, which in turn lead to other actions and their consequences. For an analyst seeking to enter this process of actions leading to other actions, intentional or otherwise, intentional actions provide an effective point of entry. Intentional actions constitute an important part of the interrelated set of actions that we have taken to constitute urbanization. By their very nature, intentions are based on other desires or needs, including the need to carryout actions others have forced on a person. Wimoa is prompted by the need to join her husband. As Thompson acknowledges, intentions would themselves be spurred by wants or desires. We can begin our exploration of the elements of actions involved in urbanization by first exploring intentional actions in the processes of the urban; this will lead us to explanations of the larger set of all that happens.

Intentional urban actions

We can begin our exploration of the intentional urban actions by considering the specific actions that Wimoa intends to carry out. Some of these actions could be spurred by a larger set of all that she wants to do or be, or what Amartya Sen refers to as functionings (Sen, 2000). These functionings could range from simple needs that do not require much thought, like crossing the road to take a bus to the city, to more considered decisions to live with her husband as a family, to possibly much more ambitious desires to be an elected political leader. She could feel the need, in varying degrees, to act upon one or more of her many functionings. The underlying set of things Wimoa wants to do or be would be firmly rooted in her imagination. Strictly speaking, there can be no boundaries on this imagination. It need not even be realistic; there is nothing to stop Wimoa, while waiting for her bus to be repaired, from hoping she would be able to fly like a bird. Yet her imagination is not independent of knowledge. The very fact that she knows it is not possible to fly without the help of a machine may well contribute to the joy of imagining what it would be like if she could. More frequently, what a person wants to do or be is influenced by the knowledge that something is possible. A child seeing an aeroplane in the sky may develop an urge to be a pilot. A sensitive person who is economically comfortable in a metropolis of the developed world may, on gaining knowledge of extreme poverty in another part of the world, feel an urge to carry out a charitable act. For another more inclined to derive joy from the pain of others, and with the required software skills, there may be an urge to create a computer virus.

Wimoa's imagination, like that of any other individual, is not independent of social, cultural or other influences. Something as personal as deciding who to marry can be influenced by social norms. This would be most evident in systems of arranged marriages in India (Gupta, 1976) or Japan (Applbaum, 1995), but even where it is a personal choice there is the possibility of falling in line with social norms (Scott, 2000). Indeed, social norms can lead to actions that are not otherwise in a person's interest. Wimoa's imagination, as a married woman in a society with repressive gender relations, may lead her to want to be a "good wife" even if that means doing all the domestic work in addition to earning a living outside the home. The influence of cultural norms on individual actions can be even more substantial. Cultural norms that encourage individuals to cause pain to themselves, such as male and female circumcision, also work through the imagination of the individual. As is perhaps to be expected, imaginations are also wide open to more contemporary influences. It is hardly unknown for conformist individuals to develop an urge to carry out specific actions that they believe are consistent with what the elite would like to do or be. As Wimoa heads out to the city she may well want to change her attire to match that of elite women in the city (to the extent that her meagre resources would allow), even if her perception of what the elite in the city would be wearing is no more than the figment of an imagination influenced by advertising and gossip.

Any impression that Wimoa's actions are based entirely on all that she wants to do or be, would, however, be misleading. As the Capabilities approach notes, a person is constrained by economic, social, physical and other circumstances. She has the capability to carry out some of her functionings and not others. In Amartya Sen's formulation, a person's capability "refers to the alternative combinations of functionings that are feasible for her to achieve" (Sen, 2000, p. 75). The idea of what is feasible could be interpreted in a very wide sense to include the constraints placed by social and even physical coercion. But when the coercion reaches a point where a person is forced to act in ways that go completely against what she wants to do or be, presenting her actions as a choice would not be accurate. This could be true of Wimoa's actions in the processes of agglomeration and polarization. Her role in these processes could be the result of the coercions of patriarchy. She may have had little option but to follow her husband's instructions, especially if the faintest signs of independence could lead to physical violence. Beyond the family, the physical coercion of gender relations in public spaces can contribute to making cities unsafe for women. The process of polarization has further elements of coercion. It can lead to extreme forms of segregation within cities, especially when identity has been used as social capital when migrating to the city. Having come to the city as a part of a group with a shared social identity there would be a tendency, at least in the initial years after migrating to the city, to function within the same social group. In a city in which there are tensions between different social groups Wimoa may need to stay away from antagonistic groups. Thus rather than seeing Wimoa's actions as being based on her functionings alone, we would be better served by a broader conception of the urge for her to act. This urge would include all that she wants to do or be, as well as all that she is forced to do or be. If she is fortunate the two would coincide. She may also come to believe that what she is forced to do is what she wants to do. Yet these two dimensions need not always be consistent with each other. When what she is forced to do or be runs contrary to what she wants to do or be, her actions would reflect the net effects of this divergence of interests. This combination of a person's inclinations and what she is forced to do or be are typically continuously changing. Some of the urges to act, like circumcision, may have ancient origins and be relatively less inclined to change while others, like what the elite is wearing, could change very rapidly. The urge to act could also vary quite substantially across individuals, some more than others. There are some urges that are closer to being universal than others. The desire to live is so widespread that it can sometimes be considered universal, though there are persons who may be suicidal. At the other end of the spectrum, there are urges that are idiosyncratic.

Very often a person may not even know why she aspires for a particular goal. She may aspire for something simply because it seems to be worth aspiring for; she may have been told it is worth aspiring for; or in some cases she might even have been forced by circumstances to aspire to be or do something. Wimoa may aspire to dress like the more fashionable women she has seen on urban signboards because it seems a worthwhile aspiration, she may aspire to be a "good wife" because she had been told that is what she should be, or she may have been pushed by her husband to aspire for a job in a garment factory. These forms of thinking have more to do with intuition rather than explicit rational thought. Thus even as it is possible for a person's aspirations to emerge from her use of cold rational thought processes, it is as, or perhaps more, likely that her aspirations have much to do with her intuition. And to complicate matters further she may also want to do something without any reason whatsoever. The urge to act would then cover the entire range, to use Arjun Appadurai's phrase, "from wishful thinking to thoughtful wishing" (Appadurai, 2004, p. 82).

The academic aspiration to explain this overwhelming diversity can result in the imposition of predetermined patterns on urban aspirations. Arjun Appadurai, who has arguably done the most to bring aspirations to the centre stage of urban discourse, is himself not immune to this temptation. He has tried to fit class related patterns into this diversity by emphasising the differences in the capacity to aspire between the rich and the poor. He first recognizes the diversity by pointing out that "aspirations to the good life tend to quickly dissolve into more densely local ideas about marriage, work, leisure, convenience, respectability, friendship, health, and virtue" (Appadurai, 2004, p. 68). But he goes on to emphasize the class element of this diversity by insisting that, "The more privileged in any society simply have used the map of its norms to explore the future more frequently and more realistically, and to share this knowledge with one another more routinely than their poorer and weaker neighbors" (Appadurai, 2004, p. 69).

Class does influence aspirations in ways that remain even after its economic pressures have diminished substantially. In her study of the *favelas* of Rio de Janeiro spread over more than three decades Perlman finds one of her subjects looking back on his aspirations for his children and insisting that "My greatest achievement in life is that none of my kids are on drugs, in jail, or murdered" (Perlman, 2004, p. 114). The fact that his children have gone on to do quite well in life has not changed his memory of his aspirations rooted in his class. But this prominent role of class should not lead to an underestimation of other influences. At the very least the influence of class could be mediated by that of the family, which in turn would be influenced by local cultures. The aspiration of Wimoa's husband to protect what he sees as the honour of his family by, say, ensuring his father is given due respect in a social gathering, may not always be fully understood by a similarly aged male inhabitant of a metropolitan city in the West. And then there are aspirations, such as those that emerge from love or hate, which need not be based on class.

The urge to act differentiates the concept of action from more widely held notions of agency. Agency is typically seen as the capacity to act (Schlosser, 2015). A person may have the capacity to carry out an act that she does not want to, or need to, do. Wimoa may have the capacity to run a petty shop in her village. The concept of agency would typically include this capacity to set up a small shop. But she may not want to set up a shop, and she may be under no pressure to do so. She may then have no urge to run a petty shop. Conversely, Wimoa may not have the capacity to carry out a particular act which she would very much like to. She might want to transform her village into a city but would not have the capacity to influence such processes. Her desire to transform the village would clearly not be a part of our understanding of her agency, but she may still have the urge to carry out the act. The concept of action we use here would find a place for actions she is not currently capable of carrying out. Her inability to achieve this desire would remain a thought that apart from being an action in itself, would influence her other actions. If nothing else, her unrealised desire for the urban could influence other aspects of her behaviour, thereby becoming an action that influences her other actions.

Wimoa need not be alone in her pursuit of specific intentional actions. She could be, and is more likely to be, a part of a larger set of actions of individuals acting independently or as groups. Being a part of a larger set of individuals or groups does not necessarily mean a loss of her individuality. Several aspects of her urge to act and the means she uses may be specific to her individual needs. Let us say that she decides, on arriving in the city, to immediately seek a job in a factory. She may be acutely aware of her diminutive physical stature and believe that it could lead the inspector at the factory gate to consider her to be below the legally permissible working age. She could then dress in a way that makes her look older. But just how far she should go to mislead the inspector could well be socially determined. Much would depend on what is accepted in the social circle in the city that she is now trying to find her feet in. The means she uses to follow up on her intention to find a job could well be culturally determined, with some cultures being more open to deception than others. It has been pointed out that cultural patterns extend to carrying out routine acts like following traffic rules. A study of parking tickets of diplomats in the United Nations showed that countries that ranked higher in an index of corruption tended to get more parking tickets (Fisman & Miguel, 2007). Wimoa's intentional actions in the process of urbanization would undoubtedly have individual traits, even as it is not independent of social, economic and cultural norms.

Intentional actions that constitute the processes of agglomeration and polarization, which are at the heart of the process of urbanisation, can be carried out through a variety of means. Sharing, matching, and learning, which constitute the micro-foundations of agglomeration would rely on differing strengths. Sharing would tend to rely rather more on social networks, matching on the economic relations that bring together different groups that need each other, and learning on the role of knowledge in an intentional action. Each of these means of carrying out an action is also open to the benefits of aggregation. The urban interests of a class can be better projected if the resources of a class are aggregated; the resources of capitalists as a class would typically be greater than that of a single capitalist, just as the strength of workers as a class is far greater than that of individual workers. The benefits of aggregation are also available to persons belonging to other identity groups.

The effects of aggregation are not limited to individual groups but can alter the course of the process of urbanization as a whole. Sharing, matching, and learning within and between groups can generate entirely new processes of agglomeration and polarization; processes whose impact is far greater than the sum of their individual parts. This has, arguably, been best captured by New Economic Geography. Fujita and Thisse, building on the work of Krugman and others, develop an interesting diagram of circular causation in the spatial agglomeration of firms and workers (Fujita & Thisse, 2000). It demonstrates the process through which multiple economic groups feed off each other to generate a process of agglomeration towards a particular city. We can enter the circular causation at the point where diverse firms locate in a city. This leads to a greater variety of consumer goods being produced in that location. The greater access to consumer goods enables workers to consume a wider range of goods with the same levels of nominal income. This increases the real income of workers. The possibility of a higher real income in that city attracts more workers. Since the workers are also consumers they provide a twin advantage to firms of providing both labour as well as local demand. These advantages, in turn, make it attractive for existing firms to expand and for other firms to locate in that city.

The dominant place for sharing, matching and learning, as elaborated by Duranton and Puga (2004), is evident in this circular causation. We entered this circular causation at the stage where firms in a city produce a greater number of goods providing workers with a greater variety of products. This requires matching the products produced and those that the workers, as consumers, demand. This match would be enabled by the access to the products being widely shared and workers learning about the utilization of these products. In addition to the actions that are intended to produce these products there are the products that were not intended. The manufacturer of garments in the developing world, producing for a global brand, would not intend to produce garments that do not meet the quality standards of global market. But once these export rejects are produced, the manufacturer would see it as a separate product with its own demand. These garments would be sold at prices that Wimoa and others like her would consider attractive. There is then another round of matching, sharing, and learning. There is a match generated between export rejects and the clothing of workers, sharing of the information about the availability of these garments, and some learning, even if it is grossly distorted, about Western fashion among workers and their families in the global South.

The economics of agglomeration gathers momentum from economies of scale. In a somewhat narrow accounting sense, economies of scale are built into the fact that the costs of each unit will decline when the fixed cost of a project is spread over a larger number of units. From the time of Adam Smith, though, there is recognition that economies of scale go beyond merely being able to spread the fixed costs of an economic activity more thinly across a larger number of units of output. These economies could arise from the specialization that is possible when the activity is undertaken on a large enough scale. When workers concentrate on individual tasks there are savings in time from not having to switch tasks. They also gain efficiency from repeating tasks, which is a form of learning by doing. And when specialization reduces individual tasks to simple repetitive actions they are easier to convert into processes that machines can carry out. An expansion in the role of specialization can be a prompt to set in motion the processes of agglomeration. The need for a specialized labour force can draw firms to locations where such a resource exists. As Marshall pointed out as early as the second half of the nineteenth century "a localised industry gains a great advantage from the fact that it offers a constant market for skill" (Marshall, 1890, 2013, p. 271).

Sharing, matching, and learning can also generate an agglomeration of agglomerations. Agglomerations based on individual specializations can share the need for particular resources and facilities. Firms in the cities of the global South that provide backend services for firms in the metropolises of the developed world would share their need for, say, English educated workers with firms marketing Western educational products. Workers in both sets of firms would also share their need for urban transportation. A few steps up the economic ladder, firms with diverse specializations could benefit from sharing the same airport or other urban infrastructure, just as information technology professionals could benefit from sharing a gated community with workers of the same class with other specializations. Urban centres could then grow around multiple specializations, allowing new, and possibly smaller, entrants to tap the same urban infrastructure. It is also possible that as the impetus provided by actions from one specialization ends, the actions of another specialization could take the lead. Cities in the global South that were once driven primarily by garment firms manufacturing for global brands, have at a later point of time become suppliers of basic software services.

The scope for sharing, matching, and learning between agglomerations would depend on the nature of the underlying specialization. There are activities that require workers with highly specialized skills. Developing a site to launch a vehicle to Mars requires the coordinated efforts of a number of people with specific specialized skills in a particular location. This contributes to an agglomeration of these people around the site, often leading to the creation of urban spaces. The precise choice of the site could be based on factors that are not urban in themselves, such as the technical considerations that go into deciding where to develop the launch site for the rockets. It could so happen that these sites are around already existing urban settlements, as in the case of Cape Canaveral in the United States, though it could also emerge in sparsely populated spaces. In either case the indivisibility of the technology-related activity ensures a need for agglomeration, even as the limited numbers of highly specialized professionals needed for the exercise limits the scope for sharing its location with other agglomerations.

Actions as what happens

When we divide actions into intentional actions and what happens, the latter tends to become a residual category. We treat all actions that are not intended as what happens. This approach brings with it an obvious question: whose intentions are we concerned with? If we take the intensions of anyone in any context, we would tend to veer towards the view that all actions are in some sense intentional. At the other end of the spectrum, if we take the actions as experienced by an individual there could be a lot that is happening around her that she does not intend. When we see the urban from the perspective of those who experience it we would need to take into account all that the person does not intend as what happens. Some of what happens may even be no more than the unintended consequences of her own actions. Having been married at an age well below the legally specified norm, Wimoa may still not be old enough to be employed by a factory following international norms. She may then choose to intentionally lie about her age. If the lie is credible she could get the job, but if it is not credible the person hiring at the gate may believe she is unreliable and choose not to hire her. Wimoa may take the decision on whether she gets the job or not as simply a matter of what happens.

Wimoa may also treat as what happens all that she does not expect. Some of these actions would be those that others may anticipate. In her first months in the city much of what happens could be unforeseen for her though others may have found in predictable. While she may be sensitive to the differences, and prejudices, that are deeply entrenched in her village, she would be new to the urban prejudices she now has to confront; she may be sensitive to caste prejudices in her village where everyone belonged to a single religious group, but may now have to deal with religious prejudices; she may be used to living within a single tribe in her village but have to face other identities in the city. In the process she could find herself in the midst of actions she does not anticipate. The unanticipated happenings could draw unprepared reactions from her that could, in turn, generate other reactions. It is not entirely impossible for a person who is new to the biases of the city to speak a language that draws her deep into existing divides. A person who is new to a racist society could act in ways that are totally oblivious to the existence of racial prejudice. She could mistakenly use as terms of endearment words that are considered in that society to be reflections of deep racial prejudice.

What actually happens around her in the city would also consist of actions that neither she nor anyone around her intended to occur. She may get soaked in an unexpected rain on her first day at work in the factory. In cities of the global South, what happens around her is often much more troubling than a drenching. She could find herself in the middle of violent ethnic conflict, or even war. Her own actions would, in turn, need to respond to what happens around her. In a war-torn city, the possibility of a firefight could be the major element of that urban condition. The actions needed for survivors to access water in the midst of war would require an analysis of elements such as the reservoirs and pipelines that remain available in that city. At the same time, those seeking routes to escape the killing would be more interested in the forms of transportation that are available. Equally, in a city far away from large-scale organized armed conflict, war would be a far less relevant factor to consider. The actions that need to be analysed would then vary from situation to situation, as would the relevant influences on those actions.

Built into this situational dimension of the analysis is a sensitivity to change. The elements that are crucial to understanding one action may be quite irrelevant to another. This could be true even of actions of individuals within the same family. The religiosity that leads one sibling from a village in south India to carry out the actions needed to become a nun in the Vatican may be completely irrelevant to the actions that drive another sibling to seek her fortunes in Mumbai. The consequences of an action could also vary across place and time. In the 1970s the celebration in Western fashion of denim jeans (Steele, 1997) and other clothes of the working class propelled the forces of agglomeration around cities in the developing world where these garments came to be manufactured. Yet denim jeans were little more than a dress code in the later processes of agglomeration spurred by the information technology revolution.

The situational dimension of the urban processes demands that each process of agglomeration and polarization, and their consequences, be understood separately. This is not to suggest that each of these processes are isolated and unconnected with each other. On the contrary, they are likely to interact quite widely, including sharing, matching, and learning from each other. But conceptualizing them separately would allow for each process to be understood in its particular context. The urban can then be seen to consist of multiple processes of agglomeration, polarization, and their consequences, occurring simultaneously. Indeed, these processes typically occur alongside other processes working in the opposite direction. Even as Wimoa and her husband are drawn to the city by the opportunities it offers for work, the conditions in which they have to survive in the city – including the severe shortage of basic facilities – may be daunting. The expressed needs of a manufacturer may draw workers to the city even as the actions of an inefficient and insensitive planner may make it difficult for them to remain in that urban situation.

Individual processes of agglomeration and polarization need not be long-lasting, though their consequences may well be. This is particularly true of agglomerations propelled by social conflict. A particularly brutal example of self-reinforcing migration to specific cities that was not prompted entirely by economic considerations would be the agglomeration that emerged from the partition of 1947 that created Pakistan out of an undivided India. The chaotic and bloody migration of Muslims from India to Pakistan and of Hindus and Sikhs in the opposite direction involved not just moving across the new national boundaries, but also to specific urban centres in the nations they had migrated to (Khan, 2007). The choice of these centres was dependent to some degree on the support provided by the governments of the respective countries, but there was also a preference for specific cities. As the migrants of this tragedy moved in desperation to spend the rest of their lives in places they had, in some cases, never seen before, their choice of destination – to the very limited extent that they had a choice – was often based on what they shared with others in that destination. This could be a shared language as when Bengali Hindus migrating from what was then East Pakistan tended to go no further than what was then Calcutta in West Bengal. Once they moved to a destination in which they shared the most with the local population, they needed to eke out a living by meeting the demands of that population. As some of them matched their skills with local demands, others learnt from them. The possibility of such a match between migrant skills and local demands in particular locations attracted more of the forced migrants. This in turn forced the state to create facilities for the refugees, which ensured an even greater flow of the forced migrants to that destination. This process resulted in points of agglomeration within larger cities that consisted largely of refugees. This particular process of agglomeration was prompted by the social, humanitarian and political disaster of the partition of the Indian subcontinent and ended as that disaster receded into history. But the consequences of that brutal agglomeration remain, with armed conflicts between India and Pakistan spilling over into the twenty-first century.

The events that prompt agglomerations can be short-lived even in calmer times. Agglomerations can be prompted by specific political events, and do not last beyond the immediate relevance of that event. This is particularly true of political mobilizations. Large scale mobilizations around a specific political issue are typically not sustainable. Sometimes the very success of the political march may be built around elements that ensure it cannot be sustained. The Million Man March in Washington DC in 1995 was a success in that it mobilized large numbers towards a political cause. But the controversies that came with the organizer Louis Farrakhan, together with the fact that it tended to ignore the cause of African American women, limited the scope of that particular mobilization (Gay, 1998). The negotiations involved in this agglomeration were necessarily designed for the short term.

Beyond the often-angry world of political mobilization, there can be temporary but regular agglomerations prompted by social events. It is now expected that there will be large numbers congregated at Times Square in New York on the eve of every New Year. Such repeated temporary agglomerations can get more elaborate over time, involving more extensive actions of sharing, matching, and learning. Arguably the largest gathering in the world is the *Kumbhmela* that is held once every 12 years on the banks of the Hindu holy river Ganges in north India. The *Mahakumbhmela* is held every 144 years in the city of Allahabad and the last time it was held in 2013 it was expected to attract 30.5 million pilgrims on a single day (the auspicious day in the Hindu calendar, *Mauni Amawasya*) (Commissioner of Allahabad, 2013). This temporary agglomeration involves negotiations to match the actions of millions of people, the sharing of whatever resources nature, the state, and other organizations provide, and what the participants believe to be a lifetime learning experience.

The temporariness of some agglomerations brings with it a further complication: What is to be done about agglomerations that take place in areas that are rural before the temporary agglomeration takes place, and will go back to being rural when the process of agglomeration is over. The small village of Bunol in Valencia in Spain hosts the annual tomato festival, La Tomatina, which attracts large numbers of tourists (Heuvelink, 2005). For that short period of time the village is subject to considerable agglomeration, demanding individuals carry out actions that are quite different from those generated by the deep personal knowledge of each other that characterises small village communities. For that period the interaction in the village between the visitors and the local population regarding, for example, places to stay, are closer to those that occur in urban spaces. The possibility of urban processes in places that are primarily villages, is another prompt to recognize the difference between the urban and the city; a prompt that this book will respond to in a later chapter. *What needs to be emphasized here is that the urban can be seen as the set of intentional actions, towards the processes of agglomeration and polarization and their consequences, interacting with what happens to happen.*

Negotiations of the urban

The vast mass of actions that constitute the urban interact in ways that strengthen the processes of agglomeration and polarization or weaken them, accentuate their consequences or minimize them. Economists have classified the processes affecting agglomeration into centripetal and centrifugal forces going back to Colby's classic paper in the 1930s (Colby, 1933; Scitovsky, 1954). More frequently the classifications have been based on much more specific, even narrow, criteria. Some models have tended to focus on the essential character of the factors influencing the externalities related to urban processes (Kanemoto, 1980). A distinction has been made between technological externalities like spillovers and pecuniary externalities that work through market prices (Greenwald & Stiglitz, 1986). Other classifications of economic models relating to agglomeration rely on more specific inputs into the larger scenario, like transport costs (Abdel-Rahman, 1996). There are shopping models where consumers bear the transportation costs and shipping models where the firms absorb these costs and benefit from any price differentiation they offer (Fujita & Thisse, 1996). The actions that constitute the operation of the centripetal or the centrifugal forces would involve an element of negotiation. Capital would be aware of the economic processes that work against a particular city, such as higher transportation costs. Capital would also be able to strengthen its bargaining power if it had the option of investing in other cities. The terms of these negotiations would also be subject to change, as when the options of an investor increase with globalization and the hypermobility of capital.

The negotiations around labour tend to be more varied. Take the case of the highly trained manpower needed for the software industry. The bargaining power of these workers is enhanced when they have more than one employment option. Even when the other options are not explicitly discussed, the fact that they do exist will influence the negotiations. As the workers share a common training, and the cultural norms that go with the educational institutions that train them, their agglomeration could result in the emergence of a local social ethos reflecting their behaviour. This, in turn, could attract others from a similar social background and training. The negotiations that emerge in this process are clearly not confined to those between the worker and her employer, or indeed her other immediate negotiations, such as those with the owner of the house she rents. There could also be larger social negotiations between the social groups that emerge from the software industry and those that are associated with other interests in the city. When this takes the form of insider-outsider conflict, the negotiations could extend to political actions involving multiple groups in the city.

The negotiations of urban actions are also not confined to the workplace, but extend to all aspects of everyday life. The simple action of crossing a road in an Indian city can involve interesting negotiations. Metropolitan cities in India have tended to place a premium on high-speed traffic, giving the once-simple task of crossing the road some element of risk. As there are not too many walkways, and the traffic often does not obey red lights, pedestrians are left with little option but to seek other strategies to cross the road. Since darting across the road is fraught with considerable danger, the preferred option is to wait till a sufficient number of pedestrians gather. Once the number is large enough for drivers to fear the repercussions of causing them harm, the pedestrians use this power to cross the road. Just how large this group should be before the pedestrians can venture out into the road would also depend on the sense of fairness expected of drivers in each city.

To get a sense of the breadth of urban negotiations we could return to Wimoa. Since we take thoughts to be actions, her negotiations would begin with herself. The power of an idea could provide a strong motivation for Wimoa to consider an action. The freedom she feels in the city, after the social shackles of her village, may prompt her to stand up to her supervisor in the factory. This would come up against her knowledge that the supervisor could cause her to lose her job. She may then consider what her other job options are, with the availability of a large number of options working in favour of her desire to stand up to the supervisor. And she may also be influenced by what she believes to be fair. Would it be fair for her to embarrass the friend who had introduced her to the factory? How she negotiates these conflicting impulses will determine her final action of either standing up to her supervisor or accepting his authority, or some mix of the two.

Wimoa's negotiations would reflect the elements of power, options, and fairness. An obvious influence on Wimoa's ability to negotiate with other individuals would be her power – or the lack of it – relative to that of others in the negotiation. Take her intended action of seeking employment from the person hiring at the gate of another factory. Though she is clearly at a disadvantage in that negotiation, she is not without her sources of power. Some of that power may be personal. She had the economic resources to travel to the city to seek a job; something that the poorest in her village would not have had. She had the social contacts to learn about the availability of the job in the factory, and perhaps even some knowledge of the kind of person she should present herself as being. She may even have been given some indications of the political environment in which the factory operates. If the factory was operating in an environment of powerful trade unions she might have first had to find support within a union. If, on the other hand, it was an environment in which trade unions were weak, she would have had to hide whatever trade union sympathies she might have had. Knowledge of the cultural background of the person hiring too could be a useful source of power. She could emphasise what she shares with the hirer, say, language, and hide anything that she may not share, say, her caste or tribe.

Her strength in the bargaining process would also be influenced by the number and nature of options she has. In a booming city in which there is a felt shortage of labour she would have the option of a number of other factories that could be potential employers. When the city is facing an economic slowdown, these options would dwindle. The options she has would also be influenced by her personal skills. As someone who has just come to the city she may have no option but to seek to enter the factory at the level requiring the lowest skills in the manufacturing process. Over time she could develop skills that could increase her options if she were to choose to negotiate for a job at another factory gate. These skills could range from those required in the manufacturing process to wider organization related skills, including the personality traits employers are looking for.

Underlying the elements of power, and the options she chooses, there could be a sense of fairness in the work environment within which she is seeking employment. This sense of fairness, as a felt need to help someone new to the city, could result in other workers giving her basic information about the availability of jobs. Once she gets the job, it could be a similar sense of fairness that could get other more experienced workers in the factory to teach her the basic skills of her job. This sense of fairness could itself be bounded by more narrow considerations. Senior workers in a factory may only be willing to train those who they share an identity with, whether it is religion, caste, tribe or race. But the very fact that they are willing to share their knowledge would have to be attributed to a sense of fairness, howsoever constrained it may be.

Wimoa's bargaining strengths do of course come up against those of the man hiring at the gate. Much of the power of the hirer would be derived from the factory he represents. The very power to offer a job is something he is conveying on behalf of the factory. He could also add personality traits to the power he is representing. A formidable form at the gate could create an atmosphere that makes employment seekers more willing to accept bare minimum wages. The person hiring at the gate is not without options either. These options could vary between his role as a representative of the factory and his role as an individual. As a representative of the factory he would have to go by the number of those seeking work in relation to the number of jobs he has to find workers for. At a personal level he could be less interested in forcing the lowest wage possible if he were on the brink of changing jobs himself. How he manages the difference between the requirements of the factory and his own preferences would also be influenced by his sense of fairness. He may take a professional view and be loyal to the interests of the factory as long as he is employed by it, or he may decide that his last few days at the factory are the time to pay back for all the perceived injustices meted out to him at the workplace.

The negotiations outside the workplace can be even more critical to Wimoa's well-being. The many negotiations of her home, from getting one to relating to others she shares it with, would account for a considerable part of her everyday life. If we move beyond the actions of everyday life to life-threatening events, she would have much more to fear from the process of polarization than that of agglomeration. A failed negotiation in the process of agglomeration would typically involve the loss of an opportunity. A negotiation at the factory gate that does not work in her favour would result in her not getting a job. This would be a severe setback, especially if her family is vulnerable. But the costs of a failed negotiation in the process of polarization can bring with it the possibility of a much more cataclysmic event. Identity groups in a city find various ways of marking out their territory in public space. In Indian cities this often takes the form of a flag or statue. A dispute over a particular public place can then result in the desecration of the symbol of the group, which in turn can generate violent reprisals. Living in vulnerable conditions in the city, Wimoa would run the risk of being the victim of such reprisals.

The extent and nature of processes of agglomeration and polarization have thus built into them varied negotiations. The disparate actors involved, seeking both similar and very different actions, point to the diversity of these negotiations, and their consequences. The vast range of the actions involved in the processes of agglomeration and polarization, together with their multifarious consequences, results in a boundless variety of urban experiences. An individual's interaction with the urban would be unique, especially in its detail. The use of actions as the unit of analysis emphasizes two aspects of urban processes that do not always get the attention they deserve: the role of ideas, and the continuous nature of urban change. Taking thought as an action brings ideas to the centre of the urban process. An idea can alter imaginations, the actions that are urged by those imaginations, and hence the course of the urban process. Ideas that support discrimination can generate a pushback and increase the potential for urban conflict, just as ideas that counter discrimination can lead to a more peaceful urban condition. An approach that recognizes the vast range of urban actions is also sensitive to the change that any one of those actions can bring to the course of the processes of agglomeration and polarization, and their consequences. This road to understanding the urban would make it evident that the potential for urban change is not confined to a few primary factors but can be sparked by individual actions. Urban riots may be traced to longer-term conflicts but they are often sparked by individual actions. For an analyst of the urban the individual action may be no more than the spark that leads to a spreading fire, but for the victims the spark is as important as the longer-term causes. And the possibility of being a victim of violent urban happenings sparked by the actions of others is unlikely to be very far from Wimoa's mind.

Bibliography

Abdel-Rahman, H. M., 1996. When do Cities Specialize in Production. *Regional Science and Urban Economics*, 26(1), pp. 1–22.

Appadurai, A., 2004. The Capacity to Aspire: Culture and the Terms of Recognition. In: *Culture and Public Action*. Redwood City: Stanford University Press, pp. 59–84.

Applbaum, K. D., 1995. Marriage with the Proper Stranger: Arranged Marriage in Metropolitan Japan. *Ethnology*, 34(1), pp. 37–51.

Colby, C. C., 1933. Centrifugal and Centripetal Forces in Urban Geography. *Annals of the Association of American Geographers*, 23(1), pp. 1–20.

Commissioner of Allahabad, 2013. http://kumbhmelaallahabad.gov.in. [Online] Available at: http://kumbhmelaallahabad.gov.in/english/kumbh_at_glance.html [Accessed 23 January2017].

Davidson, D., 1980. *Essays on Actions and Events*. Oxford: Oxford University Press.

Duranton, G. & Puga, D., 2004. Micro-Foundations of Urban Agglomeration. In: *Handbook of Regional and Urban Economics*, volume 4. Amsterdam: North-Holland, p. 2063–2117.

Fisman, R. & Miguel, E., 2007. Corruption, Norms, and Legal Enforcement: Evidence from Diplomatic Parking Tickets. *Journal of Political Economy*, December, 115(6), pp. 1020–1048.

Fujita, M. & Thisse, J.-F., 1996. Economics of Agglomeration. *Journal of the Japanese and International Economies*, 10, p. 339–378.

Fujita, M. & Thisse, J.-F., 2000. The Formation of Economic Agglomerations: Old Problems and New Perspectives. In: *Economics of Cities: Theoretical Perspectives*. Cambridge: Cambridge University Press, pp. 3–73.

Gandhi, M. K., 1969. *The Collected Works of Mahatma Gandhi*, 32. New Delhi: Publications Division, Government of India.

Gay, C., 1998. Doubly Bound: The Impact of Gender and Race on the Politics of Black Women. *Political Psychology*, 19(1), pp. 169–184.

Greenwald, B. C. & Stiglitz, J. E., 1986. Externalities in Economies with Imperfect Information and Incomplete Markets. *The Quarterly Journal of Economics*, May, 101(2), pp. 229–264.

Gupta, G. R., 1976. Love, Arranged Marriage, and the Indian Social Structure. *Journal of Comparative Family Studies*, 7(1), pp. 75–85.

Heuvelink, E., 2005. *Tomatoes*. Oxfordshire, UK: CABI Publishing.

Kanemoto, Y., 1980. *Theories of Urban Externalities*. Amsterdam: North Holland.

Khan, Y., 2007. *The Great Partition: The Making of India and Pakistan*. New Haven: Yale University Press.

Marshall, A., 1890, 2013. *Principles of Economics*. Basingstoke: Palgrave Macmillan.

Perlman, J. E., 2004. Marginality: From Myth to Reality in the Favelas of Rio de Janeiro, 1969–2002. In: *Urban Informality: Transnational Perspectives from the Middle East, Latin America and South Asia*. Lanham: Lexington Books, pp. 105–146.

Schlosser, M., 2015. *Agency*. [Online] Available at: <https://plato.stanford.edu/archives/fall2015/entries/agency/> [Accessed 30 04 2017].

Scitovsky, T., 1954. Two Concepts of External Economies. *Journal of Political Economy*, April, 62(2), pp. 143–151.

Scott, E. S., 2000. Social Norms and the Legal Regulation of Marriage. *Virginia Law Review*, 86(8), pp. 1901–1970.

Sen, A., 2000. *Development as Freedom*. New Delhi: Oxford University Press.

Steele, V., 1997. Anti-Fashion: The 1970s. *Fashion Theory*, 1(3), pp. 279–295.

Thompson, M., 2010. *Life and Action*. MA: Harvard University Press.

3 Spaces and the Proximity Principle

The actions that generate the processes of agglomeration and polarization, and their consequences, can occur some distance away from the places that are usually termed urban. Changes in a village can determine the extent of agglomeration in a city. In a local Indian rural scenario the homes of the once-untouchable castes, the Dalits, are often still away from the main village. This physical distance was reinforced, especially in the years before Indian independence, by a variety of extreme forms of social exclusion, including untouchability. The socio-political movements leading to independence encouraged the process of eroding traditional systems of caste exploitation, if unevenly. After Indian independence Dalits were given greater protection of their rights, but the process of removing discrimination proved to be a slow one. While the forms of exclusion may not have been as extreme as they once were, they did not disappear altogether. At the same time Dalits in several parts of the country gained access to education. They also gained the option of relative caste anonymity that the city sometimes offered them. This mix of discrimination and opportunities drove Dalits to participate in the processes of agglomeration. Those of the Dalits who had access to education improved their chances of gaining employment in urban centres and escaping the discrimination that had not completely disappeared in their villages. The response of the migrating Dalits to urban uncertainties tended to be very different from that of migrants belonging to other castes. Dalits, especially those who owned no agricultural land, typically had little to gain from going back to their village even when conditions in the city turned adverse.

Even as the urban offered less extreme forms of discrimination against Dalits, it was clearly not a panacea. It was not unknown for urban processes of polarization to reinforce the Dalit identity. This becomes quite evident if we pause a while to explore each of the spaces of the urban experience. The physical movement of the Dalits from the village to the city would occur in what has been understood as absolute space, in the sense that it was conceptualized in the earliest conceptualizations of space, going back at least to the time of Aristotle and developed much later by Newton. It was believed it was possible to mark a point in space, in the sense that its position was absolute and unchanging. In this physical geographical sense

DOI: 10.4324/9781003196792-3

the village marks one absolute space and the city another. This is a space that can be marked by the three dimensions of length, height, and depth; ranging from these three dimensions of their home in the village to the same dimensions of the place they live in the city, to the physical dimensions of the village as a whole to that of the city as a whole. In this conception "space is a real, mind independent entity" (Huggett & Hoefer, 2017). The process of agglomeration would make it difficult for the extreme physical exclusion of the Dalits to continue in the city. Sharing, matching and learning, which are at the heart of the process of agglomeration, would require some element of physical contact, or at least a reduced physical distance. Some of this discrimination could continue into other elements of absolute space, such as the houses they are able to rent. In addition to the economic barriers they would face, there could also be social impediments. But it is very likely that the extent of this discrimination would be less daunting than the extreme conditions Dalits often faced in villages.

As the Dalits find their way through the sharing, matching, and learning of the process of agglomeration it would be clear that even as they seek to abandon the relations of the village they are developing alternative ones in the city. Being a part of the harsh and uncertain change of migration from the village to the city they may not find the time or the inclination for an intellectual discussion of space, but their experience would lead them to reject the idea of space being independent of the mind. Their experience would be more consistent with the ideas of Newton's contemporary Leibniz, who saw space in relation to all that was around it. This relation was perceived by the mind. Space for Leibniz was an ideal, "a certain order, wherein the mind conceives the application of relations" Cited in (Huggett & Hoefer, 2017). David Harvey has gone on to argue that in this conception,

> An event or a thing at a point in space cannot be understood by appeal to what exists only at that point. It depends upon everything else going on around it (much as all those who enter a room to discuss bring with them a vast array of experiential data accumulated from the world).
>
> (Harvey, 2006, p. 124)

The definition of space in terms of its relations to other things or events implies that "there is no such thing as space or time outside of the processes that define them … processes do not occur in space but define their own spatial frame" (Harvey, 2006, p. 123). For the Dalits navigating through the uncertainties of the city they would usually find themselves in the social processes their actions place them in.

A major point of employment for the Dalits in an urban centre is a construction site. They would see the site as a workplace, a place from where they earn a livelihood. The owners of that site, and what is being constructed on it would see the same site very differently. They would see it as a place where something of value is being constructed. The different ways in which the same site is viewed would suggest a relative space; a space that can draw from

Einstein's concept of relativity. Among the more accessible explanations of this concept is the one available in Stephen Hawking's popular classic, *A Brief History of Time*. To demonstrate relative space, Hawking gives the example of a ball being bounced on a train. When one bounces a ball on the exact same spot on a table in a moving train it would appear that the space where the ball is bouncing is the same. But for someone seeing that act from outside the train the ball is in fact bouncing at different points on the earth that the train has moved to over time. Einstein went on to demonstrate that space and time are related in a manner captured in the concept of spacetime. This concept added an additional dimension to the Euclidean three-dimensional space. That is to say, if a point in space can be identified through three coordinates of distance from the axes of height, length, and depth, an event that occurs at a particular time and space can be identified by adding a fourth dimension of time. A Dalit construction worker constructing a wall on a site would be well aware that the wall he is building is designed to keep him and others like him away from the site when the construction is completed.

The time dimension becomes particularly critical when we see the processes of agglomeration and polarization, and their consequences in terms of their actions. Each action occurs at a point of time. The very nature of a space in which an action occurs, and its boundaries, could change over time. In specific urban contexts these changes in space over relatively small periods of time could be quite dramatic. On the first day of 1989 Jana Natya Manch, a Left wing theatre group, was performing a street play at Jhandapur on the then periphery of the Indian capital city of Delhi. Even as the play was in progress a procession of the ruling Congress party wanted to pass through the same street space. Safdar Hashmi the director of the play requested those in charge of the procession to wait as the play would be over in a short while. Those in charge of the procession appeared to agree and move away, but they returned a little while later armed with *lathis* (heavy sticks that could be used as weapons) and attacked the performers. As the audience panicked, Safdar and an associate tried to hold off the mob to give the performers a chance to escape. Safdar's associate was beaten to death on the street and Safdar himself died the next day of the injuries he received. This brutal murder has been seen as an assault on the ideas Safdar and the Jana Natya Manch were promoting through their performance, and there is undoubtedly an element of truth in this. Yet two days after his death Jana Natya Manch returned to the same spot and completed the play. The three-dimensional Absolute space may have been identical to the one in which Safdar had been killed. But the fourth dimension of time had completely changed the nature of the space. The angry reactions against the killing of Safdar Hashmi had made a repeat of the earlier murderous action more difficult. And the immediate trigger, of Jana Natya Manch's action of performing the play at that three-dimensional space at the same time as the political party wanting to use the same absolute space to carry out the actions of a procession, no longer existed. The change in the time dimension had completely altered the nature of the space.

The fact that the same absolute space can be perceived in contrary ways by different individuals and groups generates the possibility of a multiplicity of perceptions of space. The differences in the ability to perceive space would be very substantial if, with Ernst Cassirer, we focus on the varying capabilities of individual species. At the bottom of this ranking would be those animals that can do no more than use their basic senses of hearing, sight, taste, touch, and smell. Cassirer refers to the space within these boundaries defined by the basic senses as organic space and time. The significance of organic space is evident in many cities of the global South where animals share urban space with humans. It is not entirely unusual to find a cow sitting in the middle of a street in an Indian city apparently quite unmindful of what is happening around it. As we approach the higher animals we meet with a new form of space which Cassirer terms *perceptual* space. "This space is not a simple sense datum; it is of a very complex nature, containing elements of all the different kinds of sense experience" (Cassirer, 1956, p. 64). He goes on to argue that man alone has the capacity to use the element of thought to arrive at *abstract* or *symbolic* space. Even as Cassier explains the differences in space in terms of the capacity of species he is quite clear that it is possible for a person to experience "fundamentally different *types* of spatial and temporal experience. Not all the forms of this experience are at the same level. There are lower and higher strata arranged in a certain order" (Cassirer, 1956, p. 63). Organic space and time is the lowest level, with perceptual space being above it, and abstract space at the top. Even if we are not enamoured by Cassirer's classification we cannot ignore the fact that our basic senses of hearing, sight, touch, taste, and smell do influence our perception of space. The senses of the Dalits migrating from an Indian village to a city may well be exposed to new experiences. The sounds of church bells on a Sunday morning, or that of the Aazan calling Muslims to prayer are sounds some of them may not be familiar with. The sight of fast cars, the touch of other humans in congested places, and the smell of garbage may also be quite different from what they had previously experienced. And there would the issue of their diet being looked down upon in the vegetarian quarters of an Indian city.

The exploration of spaces has taken other forms as well. Sociologists have often functioned with Henri Lefebvre's classifications of space. To use Harvey's very effective summarization of a rather dense original text:

> Lefebvre (almost certainly drawing from Cassirer) constructs his own distinctive tripartite division of material space (the space of experience and of perception open to physical touch and sensation); the representation of space (space as conceived and represented); and spaces of representation (the lived space of sensations, the imagination, emotions, and meanings incorporated into how we live day by day).
>
> (Harvey, 2006, p. 130)

Even as Lefebvre's exploration of the social production of space provides us several critical insights into perceptions of space, the value of classifying them into three categories is not always clear. The differences in the perceptions of space within each category can be very substantial. To stay with Harvey's interpretation of Lefebvre's spaces, there could be considerable differences within the lived space between, say, imaginations on the one hand, and the meanings incorporated into how we live day by day, on the other. Those who are encouraged to leave the rural by their imagination of urban life often find the meanings they have to live by in their urban settings to be very different from what they had expected. Wimoa may have been encouraged to move out of her village by a desire to leave a space that forced her to cook for a large family. On coming to the city, though, she may find that in addition to cooking for her relatively smaller urban household she is also expected to work a full time job in order to support the family. While the space of her imagination may negotiate with the space of her everyday urban life, there are a number of important differences that are blurred by forcing them into the same category.

In a possible response to the need to pay greater attention to the variation in individual categories Harvey sets out to further differentiate within each category. He takes the differentiation between Absolute space, Relative space time and Relational space time into each of Lefebvre's categories. He does so through the medium of a matrix that has Absolute space, Relative space time, and Relational space time as its rows; and Material Space (experienced space), Representations of Space (conceptualized space) and Spaces of Representation (lived space) as its columns (Harvey, 2006, p. 135). He thus moves from a three-fold classification to a nine-fold one. This does allow for a greater degree of specificity in identifying the different spaces. The experience of walls and bridges can be quite easily located in the element that emerges from the row of Absolute space and the column of Material space; and it may even be possible to go along with Harvey and see dreams as belonging to the element in the column of Spaces of Representation and the row of Relational Space.

It is not clear, though, that this ninefold classification is completely free of the tendency to blur important differences. The effort to force all experiences of space into the nine elements of Harvey's matrix can still result in ignoring boundaries that change the very nature of urban living. For instance, Harvey assumes that Cassirer's emphasis on the biologically determined dimensions of space is entirely absorbed in Lefebvre's classification. Yet anyone living in an Indian city in which the tenant renting a home may not be allowed to eat non-vegetarian food would argue that there is a clear need to retain Cassirer's emphasis on spaces determined by the basic senses, in this case socially controlled taste.

Rather than classifying spaces it may be more useful to, in keeping with the larger method used in this book, allow for the experience of an entire range of spaces. The relevant spaces would vary with the experience. A person in the midst of an action need not even be fully aware of all the spaces she is acting in. As Wimoa participates in the process of agglomeration she is unlikely to be aware that she is influencing the space of imagination of corporate strategists of global garment brands who are looking for cities in the global South with cheap female labour. She could be expected to be more aware of the spaces she is experiencing due to specific actions she consciously undertakes. Her choice of places to eat would help define her organic space just as the place she lives in would define a part of her absolute space. As the same congested absolute space she lives in takes on multiple characters in the course of a day, she would experience the impact of relative space. A space she experiences could also be defined by its relationship to others, or relational space, as when she faces her supervisor in the factory. She would be deeply involved in her lived space, from her imaginations of where her daughter would be in the future to the more mundane everyday actions. She would be aware of the conceptualised space around her when one group builds a statue in her neighbourhood while others try to bring it down. Going forward in this book we can restrict our focus to these six spaces: absolute space, relative spacetime, relational space, organic space, lived space, and conceptualized space.

Wimoa's choice of actions within these spaces could be chosen by her consciously or they could be forced on her by, say, her husband's choices or by other larger social pressures. Among the actions she personally chooses to carry out too she could adopt divergent approaches. There could be actions that allow for a simple rational choice. She might choose not to take an air-conditioned bus because she believes she can do other things with the difference in fare she would save by taking an ordinary bus. For other actions she could rely on intuition. This is quite evident when we go by what Joel Pust describes as a family of accounts that hold "that an intuition is a *sui generis* occurrent propositional attitude, variously characterized as one in which a proposition occurrently *seems* true, in which a proposition is *presented to* the subject as true, or which *pushes* the subject to believe a proposition" (Pust, 2017).

The intuition Wimoa falls back on could have one or more of these features. When in doubt she might intuitively fall back on her caste group for support because what they believe seems true; because caste has been presented to her as being strong enough to protect and help her; and/or because others have pushed her towards accepting the beliefs associated with her caste. Wimoa's dependence on intuition would be inversely dependent on how equipped she is to understanding the spaces she is forced into in the city. The greater the doubt she has about aspects of those spaces, the greater would be her reliance on intuition.

The intuitions of various people involved in the processes of urbanization would necessarily vary a great deal. The very dependence on intuition in an urban situation would vary between someone who knows that situation well enough to make a rational decision about it and those whose knowledge of the situation is so poor that they have to entirely go by their intuition. But these intuitions need not be entirely random occurrences. There could be patterns that reflect larger principles at work. Thus, even as we look at rational explanations for behaviour during the processes of agglomeration and polarization and their consequences, it is important not to ignore the possibility of people reacting intuitively to the processes of urbanization. Indeed, since the new entrants to the city, as a result of the process of agglomeration, can be expected to have a variety of doubts about urban life, intuition may even be as much of a determinant of their behaviour as rationality. And there would be much to be gained if we could identify principles that people rely on, perhaps intuitively, when falling back on intuition.

The Proximity Principle

One attempt to identify such a principle was that of the Indian political leader and thinker, Mohandas Karamchand Gandhi. As someone who led one of the largest national movements of the twentieth century, Gandhi's ideas have understandably been scrutinised primarily in terms of his influence on the course of India's politics. Yet, along the way, he came up with several conceptual innovations. While some of them, like non-violence, have been the subject of considerable attention, several of his other concepts have been largely ignored. One Gandhian concept that has received less attention than it deserves, and which is of particular interest to our search for principles of intuitive behaviour, is his concept of swadeshi. The term itself has been used very widely both in the Indian national movement and the years since Indian independence. But there is a significant difference between the popular meaning of the term and the sense in which Gandhi used it. This term is in popular discourse usually taken to mean a form of inward-looking economic protectionism. While Gandhi did campaign against foreign cloth as one of his major initiatives in the national movement, his concept of swadeshi went far beyond a plea for protectionism. As he explained to a group of Christian missionaries in 1916:

> After much thinking, I have arrived at a definition of swadeshi that perhaps best illustrates my meaning. Swadeshi is that spirit in us which restricts us to the use and service of our immediate surroundings to the exclusion of the more remote.
>
> ·   (Gandhi, 1964, p. 219)

The use of the term "that spirit in us" suggests a reference rather more to intuition than to rational choice. The generalization implicit in the term would also suggest the recognition of a common principle in the use of intuition. The immediate surroundings could be expected to have a greater influence on the three elements of the sui generis propositional attitude that can be seen to constitute intuition: the proposition seems to be true, it is presented as true or is one which the individual is pushed to believing it is true. The immediate surroundings are more likely to influence what seems to be true to Wimoa; she is also likely to be more influenced by what is presented to her as true in her immediate surroundings; and her immediate surroundings might even push her to believing something to be true.

It must be remembered that Gandhi's use of the term immediate surround-ings was not to refer to geographical surroundings alone. Given his audience at the meeting in which he presented this definition he went on explain his concept in the specific terms of his "immediate religious surroundings" (Gandhi, 1964, p. 219). And it could just as easily be used to refer to immediate cultural surroundings, immediate family, immediate economic interests, and so on. Gandhi's idea of immediate surroundings was relevant for all spaces and not absolute space alone; it was as relevant in Cassirer's organic space as it was to the territorial space that dominated the movement for Indian independence.

Where this definition of swadeshi can come up against more serious criticism would be its apparent complete exclusion of the more remote. The remote could have some influence on actions in a particular surrounding. City planners in the global South are known to pick ideas and designs from the developed world that are geographically remote and may not even be appropriate to their immediate needs. There may even be specific situations where the remote may be preferred. As sometimes happens in Indian cities, a woman may want to save her money secretly, without the knowledge of her family. Any transactions she carries out with this money would be with others as remote as possible from her immediate surroundings.

The apparent disdain for the remote in Gandhi's definition of swadeshi can be addressed, as I do here, by seeing it as a guiding principle of individual action rather than an explanation of a larger reality. As a perception leading to an action by an individual it is limited by what that person knows and cares about. A more realistic elaboration of the situation in which a person acts may involve a variety of elements that she does not know about. But to the extent that we are only interested in what prompts her actions, what is relevant is only what she knows and cares about. The possibility of Wimoa getting a job in a factory may depend on a variety of factors she knows nothing about, including decisions made in faraway command and control centres of circuits of globali-zation. But when she carries out the action of going to the factory gate to find out if a job is available all she may be concerned about is whether the person who normally hires workers is at the gate or not.

This interpretation can be made explicit by modifying Gandhi's idea of swadeshi to reflect the fact that it seeks to explain individual behaviour rather than social trends. It would be useful then to define a *Proximity Principle that restricts a person's intended actions to her intuitively perceived immediate surroundings to the exclusion of what falls beyond these boundaries in all the spaces she acts in.*

The use of the Proximity Principle to explain intuitive behaviour rather than explicitly rational decisions has its relevance for each of the terms in the defi-nition. What constitutes an immediate surrounding or the more remote are intuitively determined rather than being laid out on the basis of objectively defined boundaries. The effective boundaries of an individual's actions are determined by her acting intuitively. And there is no guarantee that these boundaries will not change from action to action and even for similar actions at different points of time. When a pedestrian carries out the sometimes tricky task of crossing a street in an Indian city, her concern is with the traffic on that street and not with the vehicles moving in a neighbouring road. For her at that point of time the street she is crossing is her immediate surroundings while the next street is remote. If in the process, though, she pauses to curse the drivers in her city, her immediate surroundings refer to the city as a whole and the behaviour of drivers in cities across the world would become remote. The only common element that must necessarily exist across all these actions is that the immediate surroundings and the more remote have to be defined relative to each other. A person's intuitive sense of her immediate surroundings simultaneously determines what would be more remote for her.

The boundaries of the immediate surroundings and the remote are also influenced by the levels of aggregation that are involved in an action; the action of crossing the street demanding one set of boundaries and that of commenting on the quality of drivers in the city demanding another set. Moving beyond the preoccupations of our pedestrian, the boundaries of the immediate surroundings and the more remote would also change at much greater levels of aggregation. For those making policy in a small town in a relatively remote part of the global South, other cities in that country would be quite remote. For a nation in Europe dealing with the European Union the immediate surroundings would be the territory within its national boundaries with what is happening in other parts of Europe being quite remote. For Europe bargaining in the World Trade Organisation the immediate surroundings are the territory and interests of all its members while the interests of those in other parts of the world are quite remote. The immediate surroundings – remote demarcation would then depend on the actions that are being contemplated or carried out. An action that relates to agglomeration from the village to the nearby town could easily find the happenings on the stage of international relations quite remote, just as migrant labour moving across national boundaries may well find international relations a part of their immediate surroundings.

In these territorial domains we need to keep in mind Saskia Sassen's distinction between fixed and mobile boundaries (Sassen, 2006). When spaces are defined by the actions of an individual, the boundaries, between the immediate surroundings and the more remote, move with her. As we noted earlier, for a pedestrian the immediate surroundings are in that street with the next street being remote, but as she moves to the next street, it is this next street that becomes her immediate surroundings while the one she was on earlier is now more remote. As long as individuals are mobile the boundaries they draw between their immediate surroundings and the more remote will also be mobile. The boundaries of the immediate surroundings and the more remote are thus continuously changing according to the actions that demand intuitive responses. The spaces in which the actions take place add further flexibility to the boundaries between the immediate surroundings and the more remote. In the imagination generated through the conceptualization of space the fear of terrorism can make activities taking place in a distant land a part of the immediate surroundings of a fearful urban citizen.

The spaces within the ever-changing boundaries of the immediate surroundings of one person could easily overlap with the immediate surroundings of another. The possibility and extent of such an overlap would depend on the autonomy of the everyday life of the persons involved. A sage whose everyday life consists of meditating alone in the Himalayas is unlikely to find the spaces of her everyday existence overlapping that of another person. As we move from the lone sage to a monastery in the hills the potential for an overlap of individual spaces increases. The potential for overlap increases further when we move on to a village. And the processes of agglomeration that form a part of urbanization substantially increase the scope for an overlap of individual spaces, with this potential being greater in metropolitan cities than in smaller towns.

It is possible to resist this overlap even in a city by seeking autonomous spaces. These autonomous spaces can be developed at the level of the individual. A person travelling in a car with the glasses rolled up gives herself a certain degree of autonomy from those in the areas she is traveling through. The spaces of her imagination need not overlap with that of the persons in the neighbourhood she is driving through. She could drive through a particularly violent part of the city without altering her relatively tranquil imagination. Yet this autonomy is not complete. Her organic space would be shared with others occupying the three-dimensional absolute space she is passing through. If she were to use her horn she would encroach into the organic space of people in the neighbourhood who may be looking for a quiet time. In this process of negotiating the overlap in the spaces of different individuals, the scope to develop autonomy would vary from person to person. As Wimoa moves into the city from her village in a bus she would have much less autonomy in terms of the spaces she has to negotiate than the woman driving the car. Wimoa could seek autonomy when sitting in the bus, wanting to do no more than sit back and enjoy a vivid imagination, but could be interrupted by others seeking to speak with her or just by the noise in the bus. And as she enters the congestion of the city her organic space would immediately be intruded by the sight, sounds, touch, smells, and tastes generated by others in the city. In all these elements of her organic space, as well as in other spaces, she would be forced to negotiate with others.

Identity and the Proximity Principle

The boundaries laid out by the Proximity Principle would determine the nature of the intended actions of each individual. It would limit the factors a person considers when deciding on an action. The action would, in turn, influence the actions of others and hence the nature of the city as a whole. Wimoa seeking a job at the factory gate could affect the prospects of the woman next to her applying for the same job. The woman next to her may be adversely affected by the loss of the job but it need not always be so. She could well learn about a better paying job elsewhere. When the boundaries for a person laid out by the Proximity Principle cut across those of others it could generate a fresh set of actions. The actions of a city are then not just the aggregation of the individual acts of its citizens working as per their boundaries defined by the Proximity Principle. They include the actions that result from the interaction between individuals in the city, each operating within the boundaries laid out for her by the Proximity Principle.

The fact that a person acts within the boundaries laid out by the Proximity Principle does not mean that she acts in isolation. The boundaries laid out for her by the Proximity Principle could itself change over time. She may learn about new elements that alter her idea of her immediate surroundings. In the course of her job she may learn about the market for the product that she helps manufacture, which could influence her actions in her workplace. She could also choose to work towards improving or extending her immediate surroundings in ways that would strengthen her ability to carry out actions. She could attempt to get over her relative powerlessness by aligning with others with whom she shares her immediate surroundings in different spaces. She could gravitate towards others who speak her language. She could feel more comfortable with those who still retain elements of the rural in the way they present themselves. The options she exercises too would be influenced by the Proximity Principle. She could look for places to eat that are closest to what she is used to; that is, within the immediate surroundings of her organic space. The Proximity Principle would also influence her sense of fairness. If she has been told in her village about the big bad ways of the city she may well gravitate towards others who share that view. She may then try to protect the view of fairness she shares with them in the spaces of their imagination, or if she is so inclined, may choose to go by what she believes are the urban norms of fairness.

Wimoa's use of the Proximity Principle need not always take her to the same group. Those who share her immediate surroundings in their imaginations of the city – of its sense of fairness – need not share her language. It may just be a group of people from different backgrounds who happen to be new entrants to the city. As she moves in with her husband the immediate surroundings of her absolute space may be filled with others with a similar economic and social background. The home is likely to be in a neighbourhood in which others have a similar economic status. The neighbourhood may also be dominated by others from a similar social background such as belonging to the same caste or race. Wimoa can fall back on different groups as those she is negotiating with change. When dealing with an issue in the neighbourhood she could identify with the local community. When seeking greater fairness from her husband she could identify with other women in her neighbourhood.

The process of deciding the group she would like to identify with would begin with her choice of action. Her urge to act would, in turn, be the result of her imagination. She may enter the city with a rural imagination but soon find that several aspects of her value system are not shared by others in the city. Her belief in the well-established social hierarchy of her village may be continuously challenged in the city. The idea of individual castes or tribes having their own spaces is likely to be rejected by several people, or groups of people, in her new urban surroundings. In her intuitive search for value systems in the city she could interact with multiple urban discourses. As her search is intuitive these discourses need not be entirely in the realm of the rational. She could be enamoured by a film star and be drawn to the values he propagates; she could be fascinated with a political leader and explore the ideas she is associated with; or she could be drawn into a language movement. She could make this choice individually or with others. As she moves in with her husband, it is very possible that her choice would be influenced by his decisions. After a while she may also be influenced by others in her new neighbourhood. Whether she makes the choice individually or with others, the particular situation would influence, if not determine, which group she identifies herself with. She could identify with a group of outsiders to the city, a caste group, a religious group, a group of workers and so on, as the situation demands.

As Wimoa finds her feet in the city her choice of groups could be influenced by any of three elements she could use in her negotiations with others, that is, the power she could claim to have access to, the options that are available to her, and the sense of fairness that she could appeal to. She could be influenced by the power the group brings to her cause. Typically this takes the form of assertion, whether it makes her more assertive or simply allows her to use the assertion of the group. In a city of the global South charged with the assertion of the majority religious group, the mere fact that she belongs to that religious group can give Wimoa power in her negotiations with those belonging to other groups. Those belonging to the threatened minority group may be disinclined to assert themselves against her. The strength of this identity group in a particular situation would itself be the result of the actions of the group designed to achieve one or both of two purposes: change the group's relations with another group or set of groups; and change the relative status of the group in society. While the first focuses on the relations between specific groups, the second relates to the overall hierarchy of groups in society. The processes of group assertion are thus necessarily in relation to other groups. The ability of a group to assert itself *vis a vis* other groups need not necessarily increase the capabilities of its individual members. While there may be cases, especially with dominant groups, where membership of the group increases the capabilities of its members, there are a number of situations where this would not be so. The victims of inter-group conflict could, in fact, find their capabilities dramatically reduced by the actions of the antagonistic group.

Wimoa could also find that being a member of a group would give her greater options. If she loses her job, she could find that others from her village, who she has stayed in touch with, could provide her information about other employment opportunities. If she faces a financial crisis she could find members of her caste more willing to lend her money on less exploitative terms. These options are not unrelated to other groups. A group with an effective social network may succeed in getting so substantial a share of available jobs that it reduces the employment opportunities for members of other groups. But it is also possible for groups to play the role of providing options independent of other groups. The ability of a group to provide an effective safety net to a migrant member depends on a number of factors, ranging from the cohesiveness of the group to the importance given to a shared history. All these factors can exist irrespective of the behaviour of other groups in that society. This inherent independent strength of a group adds to what its members can do. The effectiveness of a group in providing options typically leads to an increase in the capabilities of its members.

Wimoa's choice of groups could also be influenced by a shared sense of fairness. This sense of what is fair may be most discussed in the case of events that are considered important in the group. The person hiring at the factory gate could well find himself in a situation where he has to choose between someone who is clearly more efficient and someone who belongs to his caste. There are groups where an uncompromising focus on efficiency would be considered fair just as loyalty to caste would be considered fair in other groups. These differences in the sense of fairness across groups could extend to more routine matters of everyday urban life. There are groups in cities of the global South where it is only fair that any member of the group walks into the house of another member to watch her favourite TV programme, just as there are groups in which this action would be seen as a gross invasion of privacy. Wimoa would tend to veer towards groups that share her sense of fairness.

Once she is accepted by the specific group she would like to identify with, Wimoa would meet Kwame Anthony Appiah's three criteria for the development of a social identity: An individual must believe she belongs to a particular identity group, others must see her as belonging to that group, and there must be a discourse around that identity group (Appiah, 2005, p. 69). Wimoa's actions would reflect her desire to join the group. The reactions of others in the group will determine whether others see her as belonging to the group. But before any of that can happen there must be a discourse that would draw her to a group.

The discourses that are available to Wimoa can occupy several spaces of her experience of the city. The discourse could present itself to her through the realm of absolute space in the form of a building. A temple, a church, and a mosque all convey the existence of a religious discourse. This discourse could be strengthened by appealing to other spaces. The appeal of modernity to an urban imagination can be incorporated through the construction of a large modern building as a place of worship. The actions that are permitted by the discourse can extend to organic space by making explicit what can be eaten or touched in that place of worship. Similarly, there would be other discourses available to Wimoa around other groups she could choose to identify with. There could be discourses that are determined predominantly in the space of imagination, such as that around a particular film star. In several Indian metropolises this discourse could lead to the creation of entities in absolute space such as larger than life cardboard images of that star.

These discourses can rely on the rational or the intuitive, though it is often both. Wimoa may be intuitively attracted to what she believes a particular identity group is associated with but as she begins to justify her choice with others she could fall back on whatever elements of rationality she can command. If the group is asserting its rights on the grounds that it has been the victim in some situation, she could repeat the attempts at rational argument the group makes, no matter how weak they may appear to an outsider. In the second half of 2017 several north Indian cities were in turmoil from protests by a dominant caste, the Rajputs, against a film that they believed had hurt the sentiments of one of their queens in the fourteenth century. It was not entirely certain that the queen was real and not just a part of folklore. But once she became the symbol of Rajput identity she had to be protected against what they believed were distortions in the film. The symbolism in the discourse around the film was such that the protestors were taking to the streets in several Indian cities before they had seen the film, indeed well before the film was released. There are discourses around specific symbols that may have greater appeal to a rational mind. Dalits in India have treated statues of their leader, BR Ambedkar, as a symbol of the larger discourse on their oppression. The character of the city is deeply influenced by the discourses it offers, and hence the various identities an individual can tap. The greater the number of identities a city offers, the greater the number of groups a person can move in and out of.

The nature of these discourses is, in turn, influenced by the number of people who identify with them in particular situations. It is possible for a group to have a larger number of members. But the relevance of these numbers depends on how many of these members are willing to carry out actions related to that identity group. Most people in a city may belong to a particular religious group. A large number of the members of this group may be willing to come out and participate in a religious event, but a much smaller number may turn up for a political mobilization based on that religion. The discourse around this religious identity would then be predominantly spiritual. In contrast, if in another city dominated by the same religious group, the members are willing to come out and participate in the political event in greater numbers than for a religious event, the discourse around that identity group would be largely political.

The permanence of this discourse in a city could also vary quite substantially. Even something that is usually longstanding like a religious identity could have sub-identities that are short lived. This would be true of groups that are formed with short-term objectives, such as a group of urban pilgrims undertaking a pilgrimage together. Individual members could identify with these groups, be identified as such, and have a clearly laid out discourse around it. Yet such groups typically do not continue after the pilgrimage is over. There are other identities that are, by their very nature, short-lived. Wimoa could find herself being a part of a group of individuals who collect around a traffic accident. Having been in the city for a while Wimoa may see herself as belonging to a group of individuals who could help, others may acknowledge her as one of the group who could help, and if there is some discussion on what is to be done there can be discourse around that identity. This group can influence not just the victims of the accident but also contribute to the larger issue of how individuals in that society deal with others in distress. Yet the group itself is not likely to last for long after the accident. The fact that groups can exist for very short periods of time does not mean that all collections of individuals should be considered identity groups. A collection of individuals waiting for a traffic light to change need not meet the three conditions that Appiah sets: they may not see themselves as being part of a group, others may not recognise them as such, and there is no necessity that there is a discourse around them as they pass by each other.

The phenomenon of persons moving in and out of multiple identities calls for a different perception of groups within the city. Wimoa acting in ways that emphasise a range of identities at different points of time militates against the tendency to see groups as a collection of individuals sharing a single dominant feature. Classes have been defined in terms of individuals sharing a particular relationship with the systems of production, such as workers or capitalists. Others might prefer to recognize classes in relatively less precise economic terms, such as the middle class. There are also a variety of other dominant features that are used to define groups, including race and caste. The keyword in all such classifications is "dominant". It is nobody's case that the individuals in a group have no other feature other than class, caste or race as the case may be. But it is usually expected that the dominant feature of the group is central to understanding other actions. Class analysis expects non-economic actions, such as those related to culture, to be traced in some way to class. Those analysing other groups based on race or caste expect these features to be critical in the explanation of the actions of individuals in these groups even outside the social domain.

The relationship between the dominant feature and other elements of the situation are also usually expected to be consistent. If we are looking at discrimination between different groups it is expected that the dominant form of discrimination rules over all its other forms. Class discrimination, for instance, is typically expected to be reflected in gender discrimination, race discrimination, caste discrimination, and so on. While this may well be the case in several situations, it is by no means certain that it will always be so. It is not inconceivable for a person to be at the opposite ends of different forms of discrimination. A man may be at the receiving end of class discrimination at the workplace but be a particularly cruel harasser of his wife at home. A woman officer in government may be at the receiving end of gender discrimination at home, but she may be the one discriminating against a particular caste at the workplace. In such cases where persons find themselves at opposite ends of the processes of discrimination at different times, they can be expected to act differently in different situations. As the situations become more diverse relying on any single dominant feature of a group may not be very helpful. It may be more useful to recognize that individuals develop multiple identities that allow them to act on the basis of one identity or the other depending on the needs of the situation.

The prominent place for the situation in our exploration of the urban points to the role played by the time dimension in the processes of agglomeration and polarization and their consequences; that is the urban process. As Wimoa develops her multiple identities she can associate with one identity group at a point of time and emphasize a different identity at another time and place. She could completely identify with other workers in her factory during the day and be equally absorbed with her religion at a place of worship in the evening, moving from her identity as worker to that of a member of a religious group. More significantly, the same place can be transformed at different points of time. If she chooses to begin her day with a short prayer in a corner of her small house, that particular place could be one of worship, though after the prayer it may be turned into a kitchen. She may need to negotiate with her family members to get that space for worship to herself at a particular time of the day. The time spent at her workplace could involve other negotiations with her colleagues or her supervisor. Her walk to work would itself involve negotiations with traffic. The multiplicity of negotiations she enters into to carry out the actions she wants at different points of time points to the significance of a widely recognized aspect of urban life: time. Agglomeration by its very nature increases the number of people interacting with each other, thereby raising the prospects of the boundaries of their immediate surroundings overlapping that of each other. This calls for multiple negotiations at different points of time, bringing with it the urgency associated with urban life.

The multiplicity of situations Wimoa faces, demanding a variety of responses at different points of time, has methodological demands on those seeking to understand the working of a city. As she carries out actions that are not always consistent with each other, an analysis of her actions could throw up a number of contradictions. These perceived inconsistencies may, however, be misleading. She may be perfectly consistent in her reaction to specific situations, coming up with identical responses to similar situations. As situations change, her very situation-specific commitment would require her to choose an action that is consistent with the new situation. What would be treated as inconsistent behaviour at the level of the individual would become perfectly consistent when analysed in terms of actions in specific situations. This reconfirms one of the core ideas of this book, that when exploring the experience of urbanization we are better served by actions as our smallest unit of analysis.

The advantage of using actions remains quite significant in trying to understand the relationship between individuals and groups. When the unit of analysis is the individual, consistency would demand that the relationship between her and the group remains the same. But once we use actions as the unit of analysis she can remain consistent even when acting differently when situations change. Her choice of action, to the extent that she has a choice, could differ right to the extent of which group she chooses to associate with. She may choose to be identified with her class when seeking higher wages in her workplace, even as she falls back on her religious group when organizing a festival. To the extent, though, that she always falls back on the same option in similar situations, these divergent actions could follow a consistent pattern.

As a person explores the many spaces of the urban, she faces a series of diverse negotiations, some of which she faces individually and others as a part of an identity group. In taking the Proximity Principle into these negotiations she tries, with varying degrees of success, to find autonomous spaces for herself in the congestion of agglomeration and the potential conflicts of polarization. The positions individuals and groups adopt in these negotiations, including the means they use, influence the character of the urban process, or more precisely, urban politics.

References

Appiah, K. A., 2005. *The Ethics of Identity*. Princeton: Princeton University Press.

Cassirer, E., 1956. *An Essay on Man: An Introduction to a Philosophy of Human Culture*. New York: Doubleday Anchor Books.

Gandhi, M. K., 1964. *The Collected Works of Mahatma Gandhi*, Vol 13. New Delhi: Publications Division, Government of India.

Harvey, D., 2006. Space as a key word. In: *Spaces of Global Capitalism: Towards a theory of Uneven Geographical Development*. London: Verso, pp. 119–148.

Huggett, N. & Hoefer, C. (eds.) 2017. *Absolute and Relational Theories of Space and Motion*, [Online] Available at: https://plato.stanford.edu/archives/spr2017/entries/spacetime-theories/ [Accessed 17 December 2017].

Pust, J., 2017. *The Stanford Encyclopedia of Philosophy*. [Online] Available at: <https://plato.stanford.edu/archives/sum2017/entries/intuition/> [Accessed 31 December 2017].

Sassen, S., 2006. *Territory, Authority, Rights: From Medieval to Global Assemblages*. Princeton: Princeton University Press.

4 Negotiations of urban politics

An intuitive preference for the immediate surroundings embodied in the Proximity Principle comes up against larger urban trends, especially those related to agglomeration. The strength of a particular process of agglomeration is often taken to be reflected in its size: the larger the process of agglomeration the greater the size of the urban settlement that emerges from it. This perception is particularly well entrenched in the narrative around cities of the global South. These cities rely very heavily, if not entirely, on their population size to claim to be metropolitan. One of the elements that go into defining a metropolitan area in the 74th amendment to the Indian Constitution is that it must have a population of a million or more. Such a linear relationship between the process of agglomeration and the size of the settlement that emerges from it is not entirely beyond debate. There can be agglomerations that generate urban centres that are important in specific contexts, but are not necessarily marked by great size. Cape Canaveral is an important centre that emerges from an agglomeration of those interested in the exploration of outer space, without being particularly large in terms of its population. But to the extent that an agglomeration, by definition, implies the coming together of a mass of people it does involve an increase in size, irrespective of the absolute size of the settlement that finally emerges. The growing size would fundamentally alter the conditions in which persons within that agglomeration can exercise their preference for their immediate surroundings.

The starkness of this change is most striking in the case of workers who move from a village in the global South to a city. Wimoa would begin with the limited immediate surroundings of her rural existence. In addition to the rather confined absolute space of her village, her other spaces are also not likely to be expansive. The people and things she relates to, the perceptions of each of her five basic senses, her imagination, and what the space of her village represents would all tend to be relatively constrained. Once in the city, each of these spaces would be substantially expanded, with its effects on her perception of her immediate surroundings. She may have considered everyone in her village as a part of her immediate surroundings, but would be much less inclined to perceive everyone in the city in the same way. In the village she would expect to know all those she passes on the street, whether or not they acknowledge her, but as a part of the process of agglomeration she would come across a large number of

DOI: 10.4324/9781003196792-4

individuals in situations where they would remain anonymous to her. As she negotiates her preference for immediate surroundings in the midst of a process of agglomeration, her actions – along with those of thousands of others – would influence the very nature and course of what emerges from that agglomeration.

The negotiation of her preference for immediate surroundings with the large numbers that a process of agglomeration brings together, occurs in multiple spaces. The desire for autonomy within a congested absolute space can be achieved, to some degree, by demarcating boundaries. A person can seek to demarcate her living space within a larger, and more crowded area. It is a norm in most Western metropolises for a person to be comfortable with an apartment in a large multi-storeyed building while being equally protective about the privacy of her home. In cities of the global South the demarcation of personal absolute space may be more constrained. In most cases, the absolute space available for a dwelling would be much smaller and it would often have to be shared with several people. As Wimoa moves into the city to join her husband she may well have to share her living space with the family of a friend or relative who has hosted her spouse in the city. When they do graduate to renting a dwelling of their own, Wimoa would still have to find her own immediate surroundings within the absolute space she shares with her family. This absolute space could be demarcated functionally in terms of what the person is expected to do within the home. This demarcation could even rein-force instruments of discrimination. When the kitchen is demarcated as the woman's place within the home it comes with the discriminatory practice of her being expected to cook for the family.

The negotiation of her immediate surroundings in the midst of the conges-tion of agglomeration would extend to other spaces. The shared space within the home could be negotiated in the realm of relative spacetime, with the same absolute space taking on a different character at different times of the day. When her husband is out of the home, she could use a part of her small home to meet friends, something she cannot do when he returns. In the process of calling her friends home she could also demarcate her immediate surroundings in relational space. Wimoa would decide who would be a part of her immediate surroundings in her relational space, and who would not. The pro-cess of agglomeration would increase the possibility of her meeting a very larger number of individuals from differing social backgrounds, but she would have the option, to some degree, of deciding who would be a part of her immediate surroundings and who would not. In her initial years in the city she may tend to confine the individuals in her immediate surroundings to those who were similar to those she knew in the village, but she could over time develop other preferences. Others in the agglomeration could place a premium on their privacy in the relational space. They could use technology to create their private space in the midst of the congestion of an agglomeration. In cities of the global South it is not unusual to find streets congested with vehicles, pedestrians and animals. In the midst of this congestion, cars with dark glasses provide those within them a considerable degree of privacy. Sometimes the technologies of demarcation are less complex. A Muslim woman wearing a *burkha* when walking on the same street gives herself a substantial degree of personal privacy, even as she asserts her religious identity to the larger agglomeration.

The negotiation of immediate surroundings within the larger process of agglomeration also extends to the organic space. A person who values her privacy can switch off the rest of the world by putting on earphones to listen to her native music while in a train crowded with people from other cultures. Others may use the organic space to bring into their immediate surroundings foreign cultures that an agglomeration has exposed them to. They could develop a desire to taste a foreign dish, to smell mountain air, to feel the fur of a pet, to see Michelangelo's David, or to hear symphonies of Mozart played by a philharmonic orchestra. The specific alterations in the immediate surroundings of organic space would also be mediated by the position of the individual in the processes of agglomeration and polarization. The economic hierarchies in the processes of agglomeration and polarization would make it extremely unlikely that Wimoa would find exposure to a philharmonic orchestra as a part of her immediate surroundings. Operating within her economic and social constraints, she is more likely to allow into her immediate surroundings some of the less expensive cuisines agglomeration brings with it.

In negotiating her immediate surroundings in the process of agglomeration she is a part of, Wimoa would consider a range of options, from extreme individual privacy to shared spaces. She may want to bring into her immediate surroundings the music she enjoyed in her village, as well as other aspects of that culture. If, along with Margaret Mead (1953), we see culture as shared and learned behaviour, we can expect Wimoa to bring into her immediate surroundings others who share the same culture. In doing so, she would tend to fall back on groups in the city that provide her access to the same culture. The negotiations of her immediate surroundings would then move from the personal to an association with larger groups.

The need for the negotiations of immediate surroundings within a process of agglomeration to be carried out by groups rather than individuals is widely felt in the economic domain. A person's economic interests could see her act in ways that mobilize her class. Workers would tend towards the immediate surroundings of the working class, just as those associated with the management or ownership of an enterprise would veer toward their classes. The nature and extent of this mobilization is not independent of the specific character of a process of agglomeration. For much of the twentieth century the prominent forms of agglomeration were typically concentrated around a single workplace, usually the factory. The wages and working conditions available in that factory were negotiated through representatives of workers and those of the management in those factories. Trade unions emerged in those factories and grew to represent the interests of the working class primarily within the workplace. While the interests of unions may have been concentrated within the workplace, the idea of class as the basis of a person's immediate surroundings spilled over outside the workplace into other aspects of everyday life. The workers could usually afford to live only in working class areas, and develop common tastes in a variety realms including popular culture. The resultant shared and learned behaviour determined working class culture. Correspondingly, other classes acted upon the impulses provided by their cultures. As each class developed its own culture, influenced by its position within the process of agglomeration, there emerged a multiplicity of urban cultures, each of which determined its own immediate surroundings.

The precise role class plays in the emergence of local urban cultures varies across processes of agglomeration. Agglomerations that emphasize the workplace bring class to the centre of much of what they do. The survival of the workplace, such as a factory and the firm that runs it, become central to the negotiations in the workplace, and the classes involved influence wider urban negotiations. Workers could be asked to consider whether the factory would survive if their demands were met. The centrality of the workplace need not, however, be the case with all agglomerations. There are agglomerations in the global South where negotiations between the representatives of capital and those of workers in a particular workplace becomes increasingly dependent on interests that are not necessarily committed to the survival of that particular workplace. The global garment industry that emerged from the fashion revolution of the sixties and seventies is an obvious example. The cultural transformation of that period saw the clothes of the working class, especially denim cloth, find a prominent place in middle and high fashion, making it possible to produce fashionable garments in the global South. This process saw the emergence of three major players: the global brands catering primarily to the buyers in the global North, and the manufacturers and the workers in the global South. In this monopsonist market, global brands exercised their power by keeping open the option of changing the source of their supplies. Specific agglomerations in the global South, built around the manufacture of a specific garment brand, could lose their engine of growth if the brand decided to shift from one source of supply to another. The uncertainty about the demand for their products ensured manufacturers in the global South kept their relationship with workers flexible, bordering on tenuous. In response, the workers, predominantly women, tried to make the best of an uncertain existence by moving from one job to another in search of higher wages. In this process the association with class organizations, especially trade unions, within the workplace was a liability. It hurt the ability of the worker to gain employment in another factory. The distancing of the worker from class based organizations in the workplace, ensured that her perception of her immediate surroundings outside the workplace was quite different from what existed within it. As the consciousness of labour and environment standards grew in the West, working conditions within factories were expected to, and usually did, maintain those standards (Mezzadri, 2012). These norms were far too expensive for workers – not too distant from the poverty line – to maintain in the immediate surroundings of their homes. Access to even these adverse living conditions often involved falling back on a worker's other identities – from caste to language to tribe – thereby moving the worker to her non-economic immediate surroundings. Class relationships within the workplace were in these cases no longer central to urban negotiations of the workers.

In some processes of agglomeration social identities could even dominate economic bargaining. In these forms of agglomeration in the global South, particularly countries like India, workers faced extreme poverty in the village, a poverty deepened by a failing agriculture. At the same time, the global aspirations of those designing cities in that country made them very expensive to live in. This made it impossible for workers, who were being forced out of agriculture, to contemplate permanent migration to the city with their families. The workers then acted in ways that helped them tap at least some of the benefit of a growing agglomeration, even as they maintained their households in the village. They worked on impermanent assignments in the cities even as they maintained their families in their villages (Haque, 2018). For these workers to get any meaningful bargain in the city they often had to present themselves as a part of a large labour group. Such groups were typically brought together on the basis of their social identity. The representative of this social identity-based group of workers bargained with employers at different locations for wages that were better than what these unskilled and semi-skilled workers could get in their villages, though they were well below what workers who were permanently resident in the city would expect. In these agglomerations, wage negotiations moved effectively from workplace based trade unions to representatives of social identities whose bargaining strength lay in their ability to move entire groups of workers from one workplace to another, sometimes in entirely different cities. Social identities can then carve out a role for themselves in the economic, social and political actions in processes of agglomeration.

The different economic patterns in each form of agglomeration ensure there is polarization between them. The actions of workers who move permanently into the city to a relatively secure job are carried out in the hope of creating immediate surroundings that can survive and even grow in the long term. Their sense of permanence may lead them to pursue longer-term aspirations and maybe even transfer their unfulfilled aspirations to their children. In contrast, workers who come as a group to carry out specific tasks in the city, say in the construction industry, have no such sense of permanence. They expect to leave the city the moment their task is done, as they would not have a place to stay once their project is completed. These groups of workers can be at, or below, the poverty line, and often move in all-male groups having left their families behind in their villages. The difference between their immediate surroundings and those of even the poorer among the workers who are permanent residents of the city, can be quite stark and polarizing.

The coming to the city of groups of workers as a part of varying processes of agglomeration provides the first round of polarization. The critical role played by social identities in the formation of these groups adds a second round of polarization. Workers from different processes of agglomeration who share a particular social identity can seek to create a larger sense of their immediate social surroundings. This would involve actions in multiple spaces, from living in the same locality in absolute space to sharing a similar imagination of their everyday life in the city. As long as each social identity can mark out its own space and remain within it, the autonomy of their immediate surroundings would ensure they do not come into conflict with other social identities. But when the actions of one or the other social identity move beyond the autonomous spaces of its immediate surroundings it could lead to conflict. These conflicts can occur in the absolute space, typically over land, but are by no means confined to that space. The effort of identity groups to protect, and sometimes grow, their immediate surroundings can extend to the representations of space, or what David Harvey terms conceptualized space (Harvey, 2006). Identity groups in a city can mark out specific spaces with their icons and claim them as their own. This is often done by placing statues of their icons at specific prominent places. In Indian cities, Dalits typically identify themselves with their iconic leader, Ambedkar, and erect statues of him at prominent places, just as other groups place statues of their own icons (Jaoul, 2006).

Globalization and agglomeration

The negotiations of the preference for immediate surroundings within agglomerations by both individuals and groups are further complicated in the twenty-first century urban context by the effects of globalization. The processes of globalization have altered both the nature of agglomerations as well as perceptions of immediate surroundings. Understanding these effects requires us to pause a brief while and outline what exactly we mean by globalization.

As is perhaps to be expected from a phenomenon that has affected so many, there is no dearth of definitions of globalization. These definitions are very often focused on the effects of the process, and sometimes only those effects that are of interest to a specific discipline. Economists tend to see globalization primarily in terms of cross-border economic processes (Stiglitz, 2012), just as sociologists tend to emphasise global social relations (Giddens, 1990, p. 64). Yet the widespread effects of globalization demand a more comprehensive and consistent view of this process; a conceptualization of globalization that can be used across multiple domains. While this may appear to be a near-impossible task if we begin with the effects of globalization, the task does become somewhat easier if we shift our focus to the causes of globalization.

Underlying the varied effects of globalization is the fact that they are caused by a reduction in the effects of distance. This reduction is invariably enabled by technological change, such as the communication revolution in the current phase of globalization. Even as technological change is a necessary condition, though, it does not ensure globalization. It is quite possible for technology to have the capacity to connect remote parts of Africa with equally remote parts of India. But without any economic, social or other reason to make that connection, the technology would not be utilised for that purpose. The realization of a particular connection of globalization would then require both conditions to be met: there must be a technology that enables a reduction in the effects of distance, and there must be economic, social or other benefits in using that technology. Globalization can then be seen as a series of links between different places – enabled by technology and motivated by economic, social or other factors – that reduce the effects of distance.

Such a view of globalization draws on Saskia Sassen's unpacking of the global economy "into a variety of highly specialized cross-border circuits corresponding to specific industries, more precisely, those components of industries operating across borders" (Sassen, 2001, p. 347). The restriction of the analysis to circuits that operate across national borders, however, misses out on some major consequences of a reduction in the effects of distance. The reduction in the effects of distance may be most dramatic across national boundaries, but they can be equally significant within nations as well, especially large nations. The communication technologies that spurred globalization could be used to tap human resources in different parts of a country. Just as communication technologies enable a firm in the developed world to tap human resources from an information technology campus in the developing world, communication technologies could also be used by a firm in a major resource centre to tap human resources from another point within that country. The effects of distance in the processes of agglomeration in the global South are reduced when contractors use mobile phones to mobilize workers from distant rural areas. In this process of agglomeration a circuit links the point where a contractor has struck a deal with an employer to distant villages that provide the workers. The term circuit is used here to refer to the entire path from the source of its inputs to the end-user of a specific product or service and back to the source, at the behest of a command-and-control centre.

The processes of agglomeration would use a multiplicity of such circuits, both within a country and across national borders. Take the case of an information technology product or service of a global brand. The command and control centre of that brand is likely to be located close to the primary market of the product or its source of finances, so as to respond quickly to fluctuating market trends and changing financial situations. The technical manpower would be located in areas that are able to provide this resource at a lower cost, typically across national boundaries. This necessitates a cross border circuit that links the command and control centre to the resource centre (Pani, 2009). The resource centre would, in turn, need to tap less skilled workers to, say, build and maintain its campus. These workers may need to be mobilized from distant places for specific tasks. This creates another circuit where those providing technical manpower to global brands are now the command and control centre and the workers in distant villages are the resources. These resources are mobilized through a mixture of information technology, which helps inform workers about the availability of the jobs, and transportation technology which brings them to the site. These workers would typically be from the same country where the campus they have to work on is located, but with porous borders even that need not be the case. We then have multiple circuits linked to each other in the process of agglomeration. The global circuit links the command and control centre to the resource of technical manpower, usually across national boundaries. The resource centres of technical manpower, in turn, become the command and control centres of circuits tapping less skilled manpower, usually within the same country.

The multiple circuits of globalization can cause a separation within the larger process of agglomeration. This is best seen in the circular causation of the process of agglomeration we had borrowed from New Economic Geography in Chapter 2. For this process to work in its entirety in, say, the production of information technology services, firms would have to move towards the locations where the technical manpower is available. The availability of jobs with these firms, and the variety of products (including software products) available in these locations would attract more technical manpower, which would, in turn, attract more firms. This circular causation has been transformed by the communications revolution. It is no longer necessary for firms to be located in the regions where technical manpower is available. Communication technology can be used to tap this manpower from a distant location. The process of agglomeration then works around two poles. One pole is the command and control centre. Its demand for financial resources, and the associated highly qualified manpower, draws those with the required skills to these centres. The availability of these skills draws other command and control centres to these locations, bringing with them their access to financial capital. The availability of financial capital, and the lifestyles that go with it, attracts others with the skills that the financial centre needs, thereby attracting more command and control centres. The second pole is where the technical manpower agglomerates. Firms that provide the resources to the global circuits attract technical manpower primarily by offering higher wages than those of competing jobs in that local economy. To the extent that the wages offered remain much lower than what would have to be paid at the location of the command and control centres, it does not affect the viability of the global circuit. The availability of technical manpower at this location attracts other firms that provide resources to the global command and control centres. The coming of these firms increases opportunities that attract more manpower. These locations also set off the secondary circuit where they attract less skilled labour to develop and maintain their campuses.

Globalization thus sets off two distinct processes of agglomeration. One process of agglomeration takes place around command and control centres. The other process of agglomeration is around the centres providing resources, both technical manpower and the less skilled manpower required to maintain the technical manpower. The two processes of agglomeration sparked by globalization have two very different sets of consequences.

The agglomeration around the command and control centres of the circuits of globalization are, in a sense, capital led. It is not just that the hypermobility of globalized capital allows it to flow into, and concentrate in, these centres. Capital is also the enabler of the agglomeration. It offers the levels of remuneration that attract individuals and firms that can manage that capital. The command and control centres also decide the products to be launched, thus influencing the consumer decisions of professionals joining this agglomeration. The decisions taken by the command and control centres of globalization also have a polarizing influence. As they move jobs from the vicinity of their absolute space to more distant lands, workers living closer to command and control centres could lose their jobs. This causes polarization between those who benefit from globalization and those who lose, or at least do not benefit, from it. The precise balance between those who benefit from being a part of the capital-led agglomeration and those who are left behind in globalization would vary from city to city.

In contrast, the agglomeration around the resource-providing end of global circuits would tend to be labour led. There are two sets of labour, with very different skill and wage levels, that agglomerate towards these centres. There is the technical manpower which is paid wages that are, though lower than those in the global command and control centres, well above those traditionally paid in the resource centre. As firms compete with each other to attract more technical manpower these workers could find opportunities to raise their wages by moving from one job to another. This attracts more technical manpower to this point of agglomeration, thereby further improving its ability to provide software and other services to the command and control centres of the global circuits. The other set of semi-skilled and unskilled labour operates at much lower wages. The much lower levels of skill of this labour are tapped by firms for more rudimentary tasks like building and maintaining software campuses. These wages are kept down by the fact that the only other option for this labour is even worse paid jobs in the villages, if that.

The process of individuals and groups finding space for their immediate surroundings within the agglomerations and polarizations generated by global circuits is further transformed by the impact of globalization. Globalization, by its very nature, alters the immediate surroundings of individuals and groups. Even as distance in absolute terms remains unaltered, the effects of that distance are fundamentally reconstructed; while the absolute space remains unchanged, other spaces are transformed.

The impact on relative spacetime is, arguably, the most widely utilized. Communication technologies have enabled the space in front of a computer or another screen to change from time to time, from being part of a cozy home, to a site of an international conference, to a more mundane official workplace, to a space for leisure, and more. These transformations of the same absolute space across time have been institutionalized through practices like work-from-home. The isolation demanded by the Covid 19 pandemic resulted in the rapid expansion of this practice. The continuously mutating relative space allows a person to alter her immediate surroundings without having to move in absolute space. This has its impact on the nature of agglomerations and consequently on our perception of the urban. Some parts of the process of agglomeration can now occur without physical movement to a single location. This enables individuals to retain some of the preference for their immediate surroundings without necessarily losing their place in the processes of agglomeration. It must be stressed that this ability to alter agglomerations in ways that allow for the primacy of immediate surroundings is not without its limits. There are some agglomerations that are less susceptible to being shifted to the virtual world. The circuit that involves accessing unskilled labour in villages for construction work in the city would necessarily require the physical movement of workers to cities. But the impact of the technologies of globalization on relative spacetime substantially expands the scope to prefer immediate surroundings within the processes of agglomeration.

The impact of globalization on the working of the Proximity Principle, and hence the scope for immediate surroundings in agglomerations, extends deep into relational space. Communication technologies, even more than transportation technologies, have shrunk the distance between points and persons across the globe. This strengthens networks that strengthen agglomeration. In circuits at the lower end of the economic hierarchy, it enables networks that allow cities to tap rural labour. Towards the higher end of the economic hierarchy it enables cross-border networks to tap labour into global circuits. The working of the Proximity Principle in global circuits is not confined to economic processes. It can also increase the scope for immediate surroundings in the social domain. Match-making sites in India enable arranged marriages between persons in cities in different parts of the world. The vast amounts of information collected on those sites allow for matches between persons of very specific and narrow social backgrounds, such as sub-castes.

The ability to prefer relatively narrow immediate surroundings while being a part of a global circuit is enhanced by the effects of communication technology on two of the five senses in the organic space. The technologies of globalization make it possible for persons situated across the globe to see and hear each other in real time. These real time visual conferences allow for a near real experience of events taking place at a physical distance, ranging from personal celebrations, like a marriage, to larger displays of joy over a football victory. The ability to jointly experience events across the globe allows social identities to spread across local and national boundaries. Communication technologies strengthen the ability of a person to see herself as a part of a group even when she is physically located permanently on the other side of the globe. Greater connectivity also enables others to recognize her as part of that group. The greater levels of communication between such individuals also creates a discourse around the identity. Thus all three of Appiah's conditions of a social identity are strengthened by globalization: it makes it easier for a person to believe she belongs to a particular group, easier for others to see her as belonging to that group, and easier for a discourse to emerge around that group (Appiah, 2005).

The globalization of social conflict leaves its mark on urban conflict. The polarization that goes along with agglomeration brings with it the scope for conflict. Conflicts have been understood in varied ways. It is sometimes associated with its causes (as in water conflicts) or with its participants (as in ethnic conflicts) and sometimes with its methods (as in terrorist conflicts). The varying perceptions of conflict have contributed to the range of its definitions. The more generalized definitions recognize a central role for a divergence of interests. Not all divergences of interests need, however, lead to conflict. It is possible, especially in stable unequal societies, for a divergence of interests between groups to be socially accepted to a point where it is not challenged. This has led to the view that "conflicts exist when divergent interests lead to incompatible actions" (Ajay, 2020). If we see conflict as a divergence of interests along with incompatible actions, urban polarization generates the divergence of interests, though it need not always prompt incompatible actions. There are thus situations that lead to urban conflict and those that remain at only a divergence of interests. These disparate propensities for conflict can be seen in both the capital-led and labour-led agglomerations that emerge from globalization. As we have noted, in cities of capital-led agglomeration there is scope for polarization between those who are, directly or indirectly, a part of global circuits and those who are not. Where the number of those left behind is very large, it is not impossible for the polarization to develop into anger and spill onto the streets. If we add to this situation the competition for the limited jobs available to them, there could also be a tendency for them to be divided on the basis of race, religion or any other identity. In cities where the number of those benefiting from global circuits is much larger than those left behind, the scope for significant incompatible actions, and hence conflict, is much less.

Polarization in labour-led agglomerations could be quite sharp, leading to a substantial divergence of economic interests between those who are a part of global circuits and those who are not. The technical manpower who benefit from globalization typically earn higher salaries than those with the same qualifications who are not a part of global circuits. These salaries are also many times that of unskilled workers. There is also a vast economic difference between the owners of firms that agglomerate this manpower and those who are outside the circuits of globalization. But it is possible for the levels of dissatisfaction to remain low enough to prevent the divergence of interests from turning into conflict on the streets of cities. The possibility of getting a job in the circuit, even if that does not actually happen, can cap the expression of dissatisfaction with globalization. The emergence of opportunities outside the circuits of globalization, as in government jobs, would also ease the potential for conflict between those in the global circuits and those outside. The availability of options, or even the promise of options, can thus reduce dissatisfaction and temper the incompatible actions between individuals and groups, that are a part of urban conflict.

The negotiations between individuals and groups, as they seek to find space for their immediate surroundings within larger agglomerations, are thus wide-ranging. The subject of the negotiations can range from the boundaries of absolute space within a home to the larger negotiations between agglomerations to be a part of a global circuit; from the sometimes silent negotiations between a husband and wife to large public, and possibly violent, demonstrations on the streets of urban centres.

Politics of urban actions

The extent and diversity of urban negotiations places a premium on the quality of the arrangements made to carry them out. These arrangements would determine the nature of the negotiations, whether they are rule-bound or arbitrary, whether they follow norms of civility or of violence, whether their results are binding or not, and so on. The individuals involved in the negotiations, whether representing themselves or their identity groups, would also like the arrangements to protect, if not project, their interests. What is to be protected in, or gained from, the negotiations would also be subject to substantial variation, covering the challenges of domestic situations, those of the neighbourhood, those of the workplace, those of larger local and national relations, and much else. The arrangements would also need to cope with change. In times of substantial unemployment labour could be willing to accept lower wages than they would in times of labour shortages. The arrangements may also have to deal with idiosyncrasies. There could be days when a person is more generous towards others and days when she is less so. The arrangements would only work if they have some degree of acceptance among those using them. If those involved in the negotiations do not believe a set of arrangements are entirely fair to them, they could try to alter them. In the absence of a better alternative they may be forced to accept the best of a bad deal. Those involved in the negotiations could then be expected to work actively to create arrangements that are the best they can get. These actions would fit into Michael Oakeshott's view of politics as "the activity of attending to the general arrangements of a set of people whom chance or choice has brought together" (Oakeshott, 1956).

Oakeshott's definition has the benefit of comprehensiveness as it covers politics at every level from the family to the city, the nation, and beyond. This comprehensiveness is necessary to explore the politics of agglomerations that range from the local to sets of global circuits. Oakeshott's elaboration of politics would also suit the collaborative aspects of agglomeration, particularly the sharing, matching and learning involved. In this realm, facilitated by cooperation, there is place for the type of rules Oakeshott believed should be followed in the practice of politics, rules that had a prominent place for civility. This politics would extend beyond those directly benefiting from, or harmed by, the processes of agglomeration and polarization, to negotiations with the state.

The emphasis on civility in the rules Oakeshott advocated would be less suited to address the conflicts that could emerge in the processes of agglomeration, and almost certainly would emerge in the accompanying polarization. The divergence of interests could occur at multiple levels: the command and control centres of individual circuits could compete with each other to get the best resources, even as the resource centres could compete for a place in prominent global circuits. The competition between elements competing for a place in a circuit, as well as the competition between circuits, would all have the potential for conflict. The activities that arrange the collective imagination in favour of one set of circuits over another would also be an important part of the politics of agglomeration and polarization. The approaches to the practice of politics that emerge in these situations need not always be consistent with the civility-based rules suggested by Oakeshott. Thus, even when the comprehensiveness of his definition is appreciated, there is less unanimity over the rules he suggests. As Chantal Mouffe has argued, there is a need to recognise that this particular set of rules is itself "the product of a given hegemony, the expression of power relations, and that it can be challenged" (Mouffe, 1991, p. 78).

There would then be different interpretations of what these rules should be. The sets of rules, different individuals and groups in the processes of agglomeration and polarization would be comfortable with, would themselves differ. Faced with the task of dealing with the awareness of consumers in Western markets, the command and control centres would prefer a set of rules that are consistent with labour and environment standards. Factories in the global South employing labour that is forced to live in conditions that cannot afford the costs of maintaining international labour and environment standards, may be less inclined to accept expensive rules designed to meet these standards. As Moufe's puts it, "Since we are dealing with politics … there will be competing forms of identification linked to different interpretations" (Mouffe, 1991, p. 79) of what the rules should be. The rules that are to be followed in the practice of politics would themselves be subject to negotiations. We would then need to modify Oakeshott's definition and take politics to be the activity of negotiating through the instruments of power, options and appeals to fairness, the general arrangements of a set of people whom chance or choice has brought together. There would be an element of chance in the opportunities a particular individual has and a degree of choice in the opportunities she chooses to pursue. These opportunities are themselves the result of larger processes. A person would not have the chance or choice of being a part of an urban process if the agglomeration itself did not exist. The process of polarization would also demand its own arrangements. These arrangements could provide autonomy to each of the polarized groups or find other ways of managing the potential conflict. Urban politics would then be the actions generating arrangements of agglomeration, polarization and their consequences. More precisely, it would be the actions of negotiating through the instruments of power, options and appeals to fairness, the general arrangements of a set of people whom agglomeration has brought together and polarization keeps apart.

The arrangements sought by different individuals and groups involved in agglomeration, and polarization, are multidimensional. They range from diverse demands for land to a stake in the narratives of the city. The demand for land would vary quite substantially from those seeking housing for workers to those seeking large campuses with a global ethos. The unequal distribution of limited resources would, in turn, require narratives that portray these and other inequalities as the price to be paid for a larger, not always defined, good. The sources of the strength of each individual or group in the negotiations would vary. The power of the command and control centres of a circuit is primarily financial, but is enhanced by their ability to find a prominent place in friendly, material-centred imaginations. The power of workers, in contrast, is largely built on their numbers and their ability to mobilize. The options for the command and control centres lie in their ability to replace one resource centre with another, often forcing existing resource centres to fall in line. On the other side of the coin, groups of workers may sometimes have the option of shifting from one workplace to another. The appeal to fairness also varies between command and control centres and organized workers. The command and control centres would embrace a narrative that emphasizes efficiency and high growth, with the promise of a trickle down to the poorest. Organized workers would tend to emphasize greater shares of growth for the poor.

A central arena for the negotiations between diverse individuals and groups involved in the processes of agglomeration, and polarization, is the realm of rule-making. The negotiations begin with the rules of rule-making: the rules that are to be followed by individuals and groups in the process of deciding the rules for negotiation in each element of agglomeration or polarization. This would include the, often unwritten, rules that decide which groups have the social sanction to set the norms for specific negotiations and which do not. The mobilization around caste or race may be perfectly acceptable in one set of social conditions, while they are not in another. The rules for rule-making, once negotiated, are held together by social sanction. The mobilization around caste or race has social sanction in the realms where it is accepted. The pressure on these rules of rule-making occur when there is a change in conditions that brought about the original set of rules. The processes of agglomeration and polarization can generate pressure on an existing set of rules of rule-making.

Agglomerations that involve the movement of large numbers from villages to cities exert a distinct pressure on existing urban norms. The process of making rules in a village is built around local hierarchies, which are, in turn, influenced by identities in the immediate surroundings of the village. In villages where castes and tribes have significant presence, these identities often become the core of group formation and hence the basis of negotiations between groups. As workers migrate to cities, these identities, especially their local variants, can lose some of their power in a negotiation. The worker may belong to a caste or tribe that has only a marginal presence in the city. An effective process of agglomeration would, in fact, attract individuals and groups from very different backgrounds. This may contribute to the view that it is important to focus on the individual rather than the social group she comes from. The large scale movements from villages to cities in the global South can also lead to conflicts between individual rights and the rights of groups, leading to further polarization. This is quite evident in urban conflicts between those favouring affirmative action, such as reservations for backward castes in government jobs in India, and those opposed to such actions on the grounds of what they believe to be individual merit.

The pressures agglomerations exert on rules of negotiation can also come from within. A command and control centre may insist on using global standards to determine what the rules of negotiation should be. These rules can be very different from local practices. The formal negotiating practices can vary quite considerably across countries. In a global circuit individual players would find it more convenient to function with the laws of their own land. The rules of which legal system are to be followed are themselves a matter of negotiations. Agreements within and between circuits usually specify the laws of which country and court would be followed in the event of a dispute. The choice of courts, and legal rules, would itself reflect the relative strengths of the players in this negotiation.

Within these broad sets of rules there are the rules of negotiation of individual contracts, both within the process of agglomeration as well as outside. The rules of negotiation within the agglomeration are sometimes not even explicitly stated. The negotiation of a global command and control centre with a resource centre works within an understanding that the command and control centre can always seek another option. While in theory, the resource centre would also have a similar option of catering to the needs of another circuit of globalization, in practice this option does not always exist. There are also negotiations outside the process of agglomeration, with the state and its agencies, to generate rules that enable or constrain particular circuits. This would range from policy changes that allow investments into certain areas of the economy to getting access to land. These negotiations could themselves be influenced by the development of imaginations that are friendly to specific products and lifestyles.

These negotiations often reflect a demarcation between economic and political strengths. The command and control centres with the ability to provide the investments needed for agglomeration offer growth. This usually comes with the promise of a trickle down effect on levels of poverty and deprivation. Growth can also be projected as an end in itself, especially when presented as a comparison to more economically successful cities. This economic power can be leveraged to gain further access to land and natural resources and further enhance the same power. The reinforcement of economic power can be developed into a norm where the entry of capital into a resource centre is seen as an end in itself. Economic power can then directly influence both the rules of negotiation around agglomerations as well as the negotiations themselves.

The dominance of economic power, including that of the command and control centres, is, however, not beyond challenge. Extremely uneven results of negotiations that become the law can lead to large scale political mobilization against that unfair law. In less dramatic times, there can be responses that bypass the rules that emerge from the negotiations between economic power and the state. When these rules are formalized into law, bypassing them involves ignoring the rule of law. This can contribute to a divergence between legality and morality. Those who believe the negotiations have been loaded against them would believe they have a moral right to bypass the law. As this process develops roots it generates systems where the formal negotiations are bypassed by informal ones.

The bypassing of formal laws has been the focus of the very substantial literature on urban informality. The variety of situations covered in this literature provide a range of conceptions of informality. Underlying these diverse conceptions is an understanding that the formal is within the realm of state regulation and the law, and consequently the informal is beyond these norms. The relatively broader of these conceptions of informality see it as "flexibility, negotiation, or situational spontaneity that push back against established state regulations and the constraints of the law" (Boudreau & Davis, 2017, p. 155). The emphasis on the lack of regulation has resulted in even official institutions being seen as having elements of informality. As Ananya Roy has argued, "India's planning regime is itself an informalized entity, one that is a state of deregulation, ambiguity, and exception" (Roy, 2009, p. 76).

The focus on the state and the law is undoubtedly an important part of the story, but it leaves out a vast set of rules that are framed outside the realm of the state and the law, and yet cover a vast mass of urban actions. This is quite evident in some personal actions, especially those relating to marriage. The state can, and does, set up norms for marriage, such as whether or not to permit same-sex marriage, but there are a variety of rules followed in the process of getting married that are not a part of state regulations. Societies have non-state routes to negotiate the rules for individuals to find partners and get married, which can be enforced with considerable force without the involvement of the state. These processes can vary from country to country and regions within each of them. This is quite evident in societies with a prominent place for arranged marriages. As in all marriages there are a variety of rules that are negotiated, usually informally. This could include rules regarding the nature and extent of dowry. There is some pressure on individual families to stay within the rules informally laid out by the larger community. Since a community is an identity group, the ability of the family to fall back on that identity for support could be eroded if it takes a course very different from that of the broader identity group. And the norms laid out for the negotiation of a dowry are themselves indirectly negotiated by the actions of its members. A member of the community could seek to establish the family's dominance with an expensive wedding, including a substantial dowry. Others who challenge the dominance of that family would need to follow similar practices, which could be evaluated in terms of the costs of the wedding. The expenditure on a wedding can then be informally used as an index of the status of the family. Once the rules for establishing status are set in such material terms, the negotiation of individual arranged marriages lower down the economic hierarchy would also have a prominent place for the material, including dowry.

This set of actions leading to marriage can be deeply affected by processes of agglomeration. A striking, and tragic, example is provided in a study of farmers' suicides in a village in the south Indian state of Kerala (Shihabudeen, et al., 2019). Cultivation in the village had become less viable, increasing the pressure to move out of agriculture. The preference for non-agricultural occupations was accentuated by opportunities available in the processes of agglomeration both within the country and outside. Work opportunities emerged in nearby cities as well as in more distant places, including in the countries of the Persian Gulf. The rural settlement saw a significant part of its population seeking to move out of agriculture, and the village itself. Cultivators could sell their land to raise the finances needed to tap such opportunities. This option was much weaker for agricultural labour. And there was a further divide within agricultural labour families. The younger men could find low-skilled opportunities outside agriculture, in, say, construction. Older agricultural labour typically no longer had the strength or stamina to make this transition. The gendered wage structure also made it difficult for young women to make the transition outside agriculture. For most agriculture labour families the future of their girl children lay in getting them married to men working outside agriculture, usually outside the village. The dowries demanded in such a marriage pushed the head of the family – the older agricultural labourer – into deep debt. And this debt sometimes led to suicide. The suicide could, in this case, be seen as the price paid by the agricultural labourer to find a place for his daughter in a larger process of agglomeration.

Once in the process of agglomeration the woman would act in ways that would help her pursue what she wants to, or is forced to, do or be. In these actions she would be influenced by her preference for her immediate surroundings in multiple spaces. This preference would influence Wimoa's choice of options as she uses whatever little power she has and appeals to fairness in each situation she faces in the process of agglomeration. As her actions pursue arrangements with those who agglomeration has brought together, Wimoa would find a place for herself in urban politics. However small this place may be, it would have room for her individuality (and the efforts of others to curb it), her association with different groups (and the corresponding potential for conflict with them), and her experience of what agglomeration can do (and cannot do) to help or hurt her.

This perception of urban politics would cover all the actions towards specific arrangements made by all who are involved in agglomeration and polarization, and the consequences of these processes. These arrangements would themselves change across situations, depending on those involved and what they pursue. Urban politics would cover the entire set of arrangements from those seeking to ease the pressures in the village, to those designed to enable the process of migration to the city, to the arrangements in the workplace and home in the city, to those further up the economic hierarchy, going all the way to the command and control centres of global circuits. The course of urban politics would then begin with the village and go on to global metropolises. Even as class, caste, tribe, race, and other groups play significant roles in different situations along the way, it would be futile to expect the actions of any one of these groups to explain urban politics at all times and across all places.

References

Ajay, A., 2020. *The Impact of Development on Social Conflicts: Case Studies of Intra-Family and Inter-Group Strife in Two Villages in Kerala*, Manipal: Manipal Academy of Higher Education.

Appiah, K. A., 2005. *The Ethics of Identity*. Princeton: Princeton University Press.

Boudreau, J. A. & Davis, D., 2017. Introduction: A Processual Approach to Informalization. *Current Sociology*, 65(2), pp. 151–166.

Giddens, A., 1990. *The Consequences of Modernity*. Cambridge: Polity Press.

Haque, J., 2018. *Land, caste and power in circular migration*, Bengaluru: National Institute of Advanced Studies.

Harvey, D., 2006. Space as a Key Word. In: *Spaces of Global Capitalism: Towards a Theory of Uneven Geographical Development*. London: Verso, pp. 119–148.

Jaoul, N., 2006. Learning the Use of Symbolic Means: Dalits, Ambedkar Statues and the State in Uttar Pradesh. *Contributions to Indian Sociology*, 40(2), pp. 175–207.

Mead, M., 1953. The Study of Culture at a Distance. In: *The Study of Culture at a Distance*. Chicago: University of Chicago Press.

Mezzadri, A., 2012. Reflections on Globalisation and Labour Standards in the Indian Garment Industry: Codes of Conduct Versus "Codes of Practice" Imposed by the Firm. *Global Labour Journal*, January, 3(1), pp. 40–62.

Mouffe, C., 1991. Democratic Citizenship and the Political Community. In: *Community at Loose Ends*. Mn: University of Minnesota Press, pp. 70–82.

Oakeshott, M., 1956. *Philosophy, Politics and Soceity*. Oxford: Basil Blackwell.

Pani, N., 2009. Resource Cities Across Phases of Globalization: Evidence from Bangalore. *Habitat International*, 33, p. 114–119.

Roy, A., 2009. Why India Cannot Plan Its Cities: Informality, Insurgence and the Idiom of Urbanization. *Planning Theory*, 8(1), pp. 76–87.

Sassen, S., 2001. *The Global City: New York, London, Tokyo*. Princeton: Princeton University Press.

Shihabudeen, P., Banerjee, D., Joshi, K. & Pani, N., 2019. *Ignored in Life, Forgotten in Death: The Nature of Agricultural Labour Suicides*. Bengaluru: National Institute of Advanced Studies.

Stiglitz, J., 2012. *Globalization and its Discontents*. New Delhi: Penguin India.

5 The place of the city

The actions Wimoa wanted to take, those that she was being forced to take, those she actually carried out, and the outcomes of those actions, including the negotiations around them, would all be associated with particular locations. These locations would usually be stationary, such as the supervisor's office, where she asked for an increase in her wage, or the point on the road where she narrowly missed being hit by a bus. There would also be times when the location was relatively mobile. When she was late for work and had to take a bus, she might have managed to find the exact same spot on the bus to stand in as she always did, but her location with respect to the world outside the bus would keep changing as the bus moved. The relative spacetime of her seat on the bus would be only one of the several spaces Wimoa experienced in the course of her actions. Despite the multiplicity of spaces she experienced, or perhaps because of this profusion, Wimoa would tend to associate each of her actions, and maybe even their outcomes, with particular locations.

These locations would usually change from action to action, though it was quite possible for the same location to be associated with a variety of actions. Each of these actions could occur in different spaces of the same location. There could be a small corner in Wimoa's home which became the location for actions in different spaces in the course of the day. It would begin the day, before the rest of the family woke up, as a relational space where Wimoa related with divinity. It would go on to become an organic space as she focused on her sense of taste when cooking for her family. It would become an absolute space when she hid her little savings in the tin containing grain, knowing that her husband would never enter the cooking area of her home. The way she arranged her meagre kitchen utensils, and the spotless character of that location, would represent her conceptualization of what her home should look like. The location would become different locales, with their own settings, during the course of the day, starting from being an area for worship, to a kitchen, to a space for hidden savings, and perhaps more. Wimoa would also expect to find different people in that location; she would not be surprised if her daughter ran in, but she would be if her husband did anything in the part of her home she considered her kitchen. That part of her home thus had a location, a locale, and a sense of place, thereby addressing the three dimensions in John Agnew's definition of place.

DOI: 10.4324/9781003196792-5

The debates on place cover too wide a range of issues to be effectively recalled here. There are issues related to territoriality, including the possibility that "localities can in a sense be present in one another" (Massey, 1994, 7), gender, identity and much else in the conceptualization of place. In the midst of these substantial and sophisticated debates, Agnew's definition has the advantage that it emphasizes place as a process rather than a rigidly fixed and unchanging location. This is evident from a closer look at the three dimensions of place he identifies.

> The first dimension is place as *location* or a site in space where an activity or object is located and which relates to other sites or locations because of interaction, movement and diffusion between them... Second is the view of place as a series of *locales* or settings where everyday-life activities take place... The third dimension is place as *sense of place* or identification with a place as a unique community, landscape, and moral order.
>
> (Agnew, 2011, 326-7)

Each of these dimensions can keep changing. The location can be mobile, as the seat in the bus Wimoa occasionally takes, and can be associated with multiple spaces during the course of a single day. The locale or settings are also subject to change, sometimes quite frequently. The settings in the place Wimoa associates with cooking would keep changing as the economic conditions of her family change. The settings of this place may also be transformed as she moves from the diet of her family in her village to that of the city. The sense of place associated with the community, landscape and moral order would also change as she and her family move from developing contacts with those in the city who have a connection with her village, to individuals and groups she had never heard of before coming to the city.

Implicit in this definition are two other dimensions of place: actions and memories. As the location, locale and the sense of place keep changing, its meaning would vary from person to person. The location where the bus Wimoa took to the city broke down, the settings including the rock she sat on, and the community of other travellers with her, all contributed to the place in Wimoa's mind. The same location and natural settings would generate a very different idea of place for the farmers in a village near that location. The very elements that ensure that a place keeps changing would also contribute to it meaning different things to different people. A major contributor to this meaning of place would be the actions a person associates with it. Wimoa would identify the place the bus broke down because of her intended action of taking the bus and the unintended one of it breaking down at that location. She could identify the place she took the bus, the place it broke down, the place she got off it, the place she had her first meal in the city, and the series of places associated with the actions that followed. Actions are thus also a dimension of place.

Not all actions would be equally relevant to a person's identification of a place. Sometimes the actions could be consistent across all locations, making it difficult for it to be identified with a particular location. Wimoa would breathe the air of a city without associating it with a particular place, unless the air was so polluted in a particular location for her to remember it. There may also be actions that fade from her memory, taking with them the places they are associated with. She may not remember the various locations where the bus stopped on the way to the city. Implicit in the identification of a place is the memory of it, often because of the actions that took place there. The choice of the actions whose location she chooses to remember would be influenced, among other things, by her current preoccupations. It would be consistent with Maurice Halbwachs' classic definition of memory as "a reconstruction of the past using data taken from the present" (Halbwachs, 1950, 1997, 119). It would also be consistent with his view that memory was associated with place, in that remembering something that happened would involve memories of where it happened.

In interpreting Halbwachs' definition to explain Wimoa's places we need to reduce, to some degree, the dependence on historians when reconstructing the past. Historians tend to be concerned primarily with events in the past that had, are believed to have had, and in some sense can be shown to have had, a lasting impact on society. Wimoa's reconstruction of her past could, however, be an entirely personal, and even private, exercise. A location could have a particular meaning for her that is not quite relevant to others. While she may or may not know about the locations that historians would like us to remember, she would have a set of locations that she remembers. These places would range from those that had a major impact on her life, both personal and social, to those, like a hoarding for a fashion garment, that she found interesting. Sometimes these memories could be functional, as when she memorizes the landmarks on her way to work in a new city. Other memories could be temporary, as when she has to remember a particular location in an unknown part of the city where she is to meet someone. The past too would be construed in a very broad sense to include even elements of the future. When Wimoa's parents visit her in the city she would go out of her way to do up her home in a way that influences how they would remember it in the future. She would plan for days to create the settings at home to provide them a memory that would give them confidence in the well-being of their daughter. Her actions would take into account how the present would be seen as the past, in the future.

The use of memory in Wimoa's construction of a place associated with agglomeration can mirror that of communities. A community can also reconstruct the past through the lens of the present. An ethnic group in a process of agglomeration can lay claim to a place using a memory it has constructed. It can claim a particular place is of religious significance to it, which can lead to conflict if another religious group makes a similar claim to the same site. It is also not unknown for places to be created in order to erase particular perceptions of a community's past. One of the, not always implicit, arguments used against India's claims to independence in the first half of the twentieth century, was that the divisiveness of its past would ensure that it could not have a modern industrial future. Among the more explicit advocates of this argument was Winston Churchill (James, 2013). To respond to this argument in terms of deeds rather than words, one of India's fledgling capitalists sought to demonstrate their ability to build a modern industrial settlement. In 1907 Jamshedji Tata began the process of raising capital for the first Indian steel plant. The nationalist dimension of the project saw it attracting capital from a large number of Indians, both the affluent and those quite distant from that economic status. The spirit of nationalism and Swadeshi that the Tatas appealed to in their prospectus ensured that there were over 11,000 shareholders in 1911 and four members of the Tata family accounted for only 13 percent of the value of all shares (Morris, 1983). The plant was located close to its sources of iron ore. The agglomeration that the plant generated created an entirely new city, Jamshedpur. If the challenge of raising funds for a new project in a poor country prevented a complete plan to be made in advance, the advantage of a Greenfield project allowed planners to contribute in stages, leading to the emergence of a well-planned city (Sinha & Singh, 2011).

Place can then be seen as a location and its settings that are associated with past, present and future actions and their outcomes for the individual and the community. Implicit in this definition are the five dimensions of place: location; locale or settings; actions; memories; and the sense of the place in terms of community, landscape, and moral order. Actions are, in several ways, the glue that binds these five dimensions together. The location of a place would be important to a person because of the actions she associates with it. The actions she takes to influence the settings (including leaving it alone if need be) determine the locale of the place. The memories are of actions and the spaces and people associated with that place. The role she identifies for the community would be influenced by what actions they take in relation to that place; the landscape would depend on how she sees that place, and possibly even alters it; and her relationship with the moral order would depend on whether it is reflected in her actions related to that place or not.

City as a collection of places

A city can be seen as a collection of contiguous places that are the destination of one or more substantial processes of agglomeration. What is to be considered a substantial enough agglomeration would vary, accounting for the wide variety of norms across the world to decide what is a city. It would not require the agglomeration of a very large number of scientists of make a knowledge city, even as other agglomerations may require larger numbers in order to be noticed. There may also be some weight given to the permanence of the agglomerations. But for those seeking to understand the dynamics of a city, the precision of the dividing line between the city and the village would not be an insurmountable barrier. If human settlements were to be arranged according to the extent of agglomeration, the ones over which there is dispute over their being a city or a village would account for a relatively small proportion of them. In a vast majority of the cases there would be agreement on those that reflect a significant enough degree of agglomeration to be better understood as being a city. A similar emphasis on being generally right would also help overcome the issue of when places are considered contiguous or separate. The challenge of deciding what places are to be considered contiguous is further complicated by the fact that contiguity could be conceived of in multiple spaces. Contiguity would usually be in the absolute space, but this need not always be the case. There could be contiguity in other spaces without contiguity in absolute space, leading to the creation of virtual cities. Whether a Google or a Facebook can be considered a virtual city may be open to debate, but there is no denying the fact that they are large agglomerations. Their association with processes of polarization is also being widely acknowledged in governance. And the consequences of these processes of agglomeration and polarization beyond the absolute space, are both wide-ranging and intense, sometimes contributing to riots (Tønnevold, 2009).

Each of the places in the collection that makes a city would bring with them their own versions of the five dimensions of place. They would have their location, locale, actions, memories, and a sense of place. The collection of places would not necessarily fit neatly into a city, as if they were pieces of a jigsaw puzzle. The places could overlap both in absolute space as well in other spaces. This overlap of places would generate both conflict and cooperation. The place Wimoa uses for cooking in her home could also be the place of her child's play, but the cooperation between the two of them would reinforce both their memories of the place. In contrast, the overlap between the place Wimoa sees as the pavement and the place the bus driver sees as the road could lead to conflict with devastating consequences. The collection of places that form a city would then not just be an aggregate of its individual parts, but would be a continuously changing consequence of the interaction between multiple places. At any point of time this collection of places would form a unique city. The collection would have its own version of the five dimensions of place. This version would reflect the various elements that have influenced the dimensions of place; individually and together.

The location of a city would be the dimension that is most affected by the processes of agglomeration. Some cities would be the sole destinations of specific processes of agglomeration, while others, especially in global circuits, would be one of two or more destinations that a process of agglomeration generates. Some cities may be dominated by a single process of agglomeration, but the places of most large cities would be the result of multiple processes of agglomeration. There could be a process of agglomeration that brings unskilled labour to the location of the city, alongside another process of agglomeration that brings in technical manpower. The precise location of these different processes of agglomeration need not be the same. Technical manpower would prefer to live and work in a part of the city which is separated from the places where the unskilled labour would live. But as long as the actions of one group have even a peripheral connection with those of another, they would tend to move at least some of their activities to contiguous places. The unskilled labour that would like to sell cigarettes to the technical manpower would have reason to set up their shops just outside the campuses of the high technology industries.

The possibility of conflict when locations overlap can at times lead to very bitter contestations. Spurred by the spurt in economic value that is brought to a piece of land by the processes of agglomeration there can often be intense conflict about ownership of that location. The greater the spurt in the price of land the more the incentive to lay claim to even the flimsiest of cases for ownership. The spurt in value can come from a variety of urban transformations. A piece of agricultural land just beyond a megacity in the emerging world could find its value increasing manifold when the city engulfs it. In situations where the village records of agricultural land are far from perfect, the newly valuable land could find many claimants. The contestation over location need not be prompted entirely by changes in the value of real estate in absolute space. The contestation could be in other spaces as well. In the relational space there could be a contestation over the location of a shop selling cigarettes near a school. Taken together the extent and nature of the contestations over location would add to the sense of the city as a whole.

The locale or settings of the city would take on both tangible and intangible forms. Among the more visible tangible settings are the architectural choices of the places of a city. This could take the form of the architecture of a particular place being added on to that of other places of a city. The places in Rome that were built during the period of fascism were characterized by large concrete blocks that emphasised size and uniformity over aesthetic innovations. The patterns that emerge from bringing multiple places together can themselves alter the settings of a city. There are cities in the United States where entire streets, and even sets of streets, have houses with very similar architectural designs in contrast to cities in South Asia where there is typically a much more anarchic combination of architectural designs on a single street. The intangible settings may be, by definition, less visible to the new entrant to the city, but for those who have lived in it for any period of time it would be no less obvious. There would be blocks in a city that would be considered unsafe by all who live in and around that neighbourhood. Places in a city could also have a cultural ethos. There could be places in a city that are associated with a particular cuisine, theatre, or mass celebration. A place could also make a statement of its culture by the sounds around it. The tangible and intangible settings of a place are often quite closely related. The high walls and small ventilator-like windows of large houses in Riyadh are designed not just to keep out the sand, but also to meet the demands for privacy in that culture, especially for women.

The settings of a city are also subject to considerable contestation. The megacities of the global South have to continuously address the challenge of accommodating those who come to it from villages in the more poverty stricken parts of their countries. The hutments that these workers build for themselves are often seen as slums by those who already reside in the city, particularly in the relatively better off places of the city. Much as this elite may need the services provided by that labour, they are unwilling to accept the settings they bring with them. There is then pressure on the city administration to move the slums to areas that are less visible to the elite. The conflict between the elite and the poor migrants can then take the form of a contradiction between the demand for services and the elite idea of the appropriate settings for the places they associate with. Within this rather extreme confrontation between the elite and slum dwellers there are multiple other contestations over the settings of place. The global circuits of the information technology industry resulted in the creation of campuses in the cities that provided the technical manpower for the circuit. These places were built on land that once belonged to villages neighbouring the city. The precise locations of these campuses were also influenced by the result of competing claims to the same land from other information technology companies as well as manufacturing units such as the ones that Wimoa would seek to work in. The locale of these places would reflect the negotiations between these individuals and groups in a city. Often these negotiations ended in local demarcations between the land of the factory and that which remained with the village. The locale of a city in such cases was an odd mix of expensive high technology campuses alongside much poorer neighbourhoods.

This mix of locales brings with it varied memories. There are the memories of older village folk of a time when what is now the city was a swathe of cultivated land. Some of them would remember that past would nostalgia, especially if the city has not treated them well. Equally there would be those among them who were exploited in the village, who might be happier to forget the past. Those who have come to the city as a part of a global circuit would like to create a locale that would be remembered as if it were a part of a Western metropolis. The settings of these places would not be determined by humans alone. As these places had in an earlier form also been home to cattle, these animals too would tend to continue using the place. This contributes to the roads in cities of South Asia finding place for a variety of animals.

The diverse memories associated with a single location can also be a basis for contestation. When memories move from the personal domain to the social, they can be associated with tangible objects. These objects could be remnants of the past, including monuments that capture specific memories of bygone eras. Where monuments of the required authenticity are not available to meet the needs of a particular reconstruction of the past, there is a tendency to create modern objects designed to remind us of the past. This could take the form of museums or statues in public places. The statues could be designed to capture memories of that place. They could also be used to bring in memories of distant place. The bust in London of the twelfth century poet-saint from south India, Basavanna, is meant to bring in a memory that is distant in terms of both geography and time.

When these memories are associated with communities they can become the basis for aggressive contestations. In some cases these contestation could just be between the thinking of a community in the present as opposed to its views in the past. The statue of a civil war hero in the American south could be associated with attitudes to race that are no longer acceptable. When the transformation in thinking is complete the community can decide whether it wants to remove the statue or keep it as a reminder of values that should not be allowed to return. But statues can also be the source of more complex associations with place. A civil war hero in the United States could mix a deplorable attitude to race with the protection of a regional identity that many in a region would want to be a part of their reconstruction of the past. In the horizontally expanding cities of South Asia the contestation over statues can take a different form. Different communities can lay claim to the newly urbanized areas by associating the place with their reconstruction of the past. This is often done by setting up statues of their heroes from the past. Indeed, there are also cases where present-day leaders cannot wait to become a part of the past and set up their own statues. The battles over the association of a community with a place could then take the form setting up, and bringing down, statues.

The attitudes to monuments would also be influenced by the nature of the elite in a city. A traditional elite with a history of its own would like to build an ethos of remembering history. This could take the form of sponsoring the maintenance of a historical monument. In contrast, a new elite may be less inclined to celebrate a city's history. An emerging elite would have reason to emphasize the present over the past. When it uses the facts of its present to reconstruct the past, many of the monuments in the city may find no place in that reconstruction. Once the new elite has built the settings of its locale in line with their view of their role, and created a community through the spaces they control, they would like to quickly develop memories of the places they have created. These memories have to be consistent with the requirements of the new elite's perception of their present and their future. As they pursue their task of creating memories of their new places, they have little reason to support contrasting memories that older heritage sites may bring.

The process of a collection of places becoming a city is thus marked by contestations in each of the dimensions of place. The contestations in the dimensions of location, locales, communities and memories are manifested in the actions they generate. The actions that are generated in contestations over location can result in a variety of characteristics that are associated with a city. The contestations over location can result in the emergence of a land mafia in a rapidly expanding city of the global South. The contestations over locale would be reflected in the impression a city creates in the mind of someone visiting it for the first time. There are cities that give the first impression of perfect order, just as there are others that wear their near-anarchy on their sleeves.

The processes of negotiation associated with the contestations over place can create places of their own. A place could reflect the generation of knowledge that forms the basis for the different positions taken in these contestations. These places could themselves be quite diverse, reflecting the many different ways in which people gain what they believe to be the knowledge required for particular actions, including those involved in the processes of contestation. These could be no more than places of gossip which provide information about where jobs are going to be available. The places that provide this first-information would have a variety of settings. They could be pubs frequented by information technology professionals or the long lines before the public water tap where Wimoa gets most of her gossip. The places of knowledge could also involve the generation of greater technical skills than mere information. Each of these places too could have very different settings. Wimoa's friends may pick up tailoring skills during the lunch break in a factory from friends who allow them to use their machines when the supervisor is not around. Others may gain knowledge from more formal schools and colleges.

The power that is used to push contestations in a particular direction is also reflected in the places powerful individuals or groups choose to be associated with. The demonstration of this power could take the form of imposing architectural structures that are designed to reflect financial muscle and exposure to particular aesthetics. The choice of locations to demonstrate this power would also be influenced by the nature of the city as well as the audience for the demonstration of that power. An elite that would like to demonstrate its power to the world at large would pride itself in living in the most expensive real estate in a major metropolis. An emerging elite would have different ambitions. New elites in the global South that have emerged by providing manpower to global circuits controlled in the West often feel a need to dissociate from the old city and build campuses, and homes, that would be comparable to places in the West. This would require land on a scale that would only be available on the periphery of an old city.

The course of the negotiations over place are also influenced by the opportunities that a city provides. A city with more options in the availability of land would have less reason to get into intense conflicts over a particular location. These opportunities could be provided in the course of everyday life in a city. In the simplest case, a child could find places in a city that provide her the opportunity to develop her sporting skills. These places could take the form of playgrounds or just basketball hoops in much smaller open areas. The places of opportunity can also take more structured forms, as in educational institutions. These institutions provide the locations for learning, the settings required for it, the community and its social capital, as well as the memories both of individuals who have been there as well as that of the place itself. The places of opportunity can emerge around other locations as well. A political rally may bring in a community of a large number of people to a particular location with its appropriate settings, such as a stage, to generate memories of the power of the person or party who organizes it. This place of power could simultaneously create places of opportunity around it. Vendors could create their own settings of little mobile carts and reach out to the same community by selling flags of the party organizing the rally.

These negotiations would bring with them several elements of uncertainty. There would be uncertainty over whether a particular urban action would finally come about. For Wimoa, there could be doubts about the safety of her child when at a playground or at home. As uncertainty mounts in several of the spaces of everyday life, a person may feel a need for places of calm in a city. This could take the form of a quiet park where it is possible for individuals to be left with their thoughts. The settings of such a park, as opposed to one designed for children to play, would ideally allow for as little disturbance as possible, leaving the persons using it with their memories. Others may find their places of calm within religious sites. Churches, mosques and temples often have places within them that are suited for meditation. Yet others may find their places of calm in a quiet corner of their homes.

The variety of contiguous places, and their contestations, can involve a multitude of actions that come into conflict with each other. A part of this conflict can be avoided by allowing places a considerable degree of autonomy. The autonomy can be enforced through physical barriers, as in a gated community or in the entrance to a sports stadium, or through practice, when those associated with one set of places believe they have little reason to go into other places. But even in cities that allow for considerable autonomy there would be common areas. The roads connecting different places would need to have common rules for traffic. It would hardly do for vehicles to travel on the left of the road in one place in a city and have to move to the right of the road in another place. When arriving at these common rules it is necessary to see the city as a single place that can generate a consistent set of enforceable norms by which the city can be governed. Governance demands that the diversity of the city as a collection of places be seen as a single uniform place which can generate a consistent set of rules.

City as a conceptualized place

The conceptualization of the city as a single unified set of contiguous places – on a large enough scale of agglomeration, living by a set of a consistent rules – would lead us to the idea of the city as a conceptualized place. This conceptualized place would provide a unified idea of its location, locale, actions, memories and its sense of place. Cutting across the vast diversity of places that form a city, this place could only emerge in the conceptual space of a city. And yet it is central to the city, in that the rules it generates extend to the other spaces of the city. This conceptualization would determine which rules are to be followed by all who are associated with the city, such as traffic rules, and which rules are to allow for variation across different places of a city, such as location-based property taxes. The distribution of places between those where uniformity is demanded and those where it is not, would vary from city to city. Some cities, like those in the Kingdom of Saudi Arabia, could demand that all places follow the solemnity of the time of prayer while other cities could confine such solemnity to its places of worship.

The task of deriving a consistent conceptualized place from the vast diversity of the city as a collection of places is complicated by the variety of conceptualizations that would be available. Each person would have her own collection of places that form the city for her, and individual interpretations of each of the dimensions of those places. Wimoa would have her own perception of the location of the places of a city. There could be places in the city she has never visited and may not know about. They would not enter her perception of the locations of the city or its settings. She would not associate those places with any actions or communities and hence would not have anything to remember it by. Conversely, her perception of the places of a city would tend to emphasize those she is most closely associated with; the places that are the site for most of her actions, typically her home, her place of work, the market where she buys her daily requirements, and the like. She would also have her own interpretation of the dimensions of these places. She would identify the location in the context of her home or her place of work. The settings she would notice would be different from that of others; she may be more sensitive to the quality of the pavement on the road than someone who drives past the place. The actions she associates with the place are more likely to be her own, and the community she associates with the place may not be the same as that of others. The variety of places in a city is thus open to unique interpretations by each of those who relate to it.

Each of the perceptions of the conceptual place of a city may be unique but they are not entirely unconnected. As individuals share experiences they would also share their perceptions of places. Wimoa is likely to share several dimensions of her place of work with other workers in her factory. She is likely to share its geographical location with other workers, especially if high walls clearly demarcate the boundaries of her place of work. When she first joins a factory she may not notice all the details of her settings the way the workers who have been there longer do, but she would share the perception of much of those settings with others. Her perception of the actions associated with the workplace may not be identical to that of others doing the same job in the factory – some may be more focussed on the job while others may use every opportunity to gossip – but they are likely to share several elements of the actions each of them associates with their workplace. They would share a sense of the community at work, even if their commitment to the moral order of the workplace does vary. And there would be similarities in their memories of their place of work.

These shared elements can be put together, in the conceptualized space, to create a conceptualized place that would, ideally, be shared by all those who are associated with a city. These conceptualisations could vary quite widely from city to city. There are cities whose conceptualization reflects a pride in building the largest buildings and the tallest towers, or what has been called the edifice complex (Sudjic, 2005). Other cities are proud of their commercial institutions, as in the claim of Peter Ackroyd (2001) that

> the events of 1986 heralded a sea-change in the position of the City of London. Its foreign exchange market is now the most advanced and elaborate in the world, handling approximately one-third of the world's

dealings; with 600,000 employed in banking and allied services it has become the largest exchange in the world.

Unlike the city as a collection of places, the conceptualized place of a city would offer a precise perception of each of the dimensions of place. Even if it does not have a clearly defined set of actions that are seen to be acceptable to the city, it would have a list of actions that are not acceptable, and may even be punishable. There would be a precise statement of its location with clearly demarcated boundaries. The settings would be clearly identified with maps that locate each of its elements, ranging from the city's monuments to its bus stations. The norms of the city as a community would be clearly laid out, with its implicit moral order. And the effort would be to ensure that this is how a city is remembered.

Conceptualizing such a place that is accepted by all in a city comes up against the varied experiences shared within different groups in a city. The experiences of the city of its elite belonging to a particular ethnic group would be very different from that of those at the opposite end of the economic spectrum, especially if they belong to another ethnic group. The presentation of the experience of the city would also vary between groups with different interests. The elite in a city in the global South that is seeking to attract foreign capital would like to highlight that part of their experience that they believe would be appreciated in the developed world. It may seek to present its own experience as that of the city as a whole, even if that requires keeping the conditions in other places of the city out of public view. A recently emerged elite may not also be too keen to draw attention to the history of the city; a history which may shift the spotlight to other groups living in that collection of places. In contrast, those at the other end of the economic spectrum may not be averse to letting the poverty ridden condition in their places in the city be known to a wider audience, if that were to get the attention of the authorities. If these groups identify themselves with an icon of the past they would also be keen to get that part of the city's history recognized as a part of the city as a conceptualized place.

The desire to conceptualize a city differently, leading to a variety of conceptualized places of the same city, is further accentuated by the fact of place being a process. The city as a collection of places would then also be a process. As a process it would be subject to continuing change, reflected in, among other things, the actions that prompt that change. Each change in the actions of a city could potentially raise questions about an existing conceptualization of the city. The city could have been conceptualized around trade, or even particular items of trade, say, a city known for spices. If a group of large manufacturers come in they may want the city to be conceptualized differently, say, a steel city. Again, if the manufacturers are seeking foreign investment in their firms they may want to present the conceptualized place of their city to be one of a global city. There is thus always tension between the city as a collection of places and the city as a conceptualized place.

The tension between the city as a collection of places and the city as a conceptualized place actively influences the course a city takes. As a collection of places, a city in the global South could provide for a variety of very different economic groups, some rich and some very poor. An elite conceptualization of the city may want to present it as a place that is attractive to global investors with all the material amenities, and the ethos, of a global city. Since transforming all the places of the city to make it consistent with the conceptualized place of a global city would be a long and arduous task, there could be an effort to hide the parts of the city that are seen by those who matter. Those who seek a conceptualized place of a global city may only be concerned with what foreign investors think. They may then work to develop the city in a way that bypasses, for foreign visitors, all parts of the city that are marked by poverty or the near absence of order. They would typically do so by ensuring the building of an airport comparable to those in the global North. They would also build campuses that can be compared to the locales of the workplaces in the West. There is then a priority to connect the modern airport to the westernised workplaces in ways that bypass the places of poverty and lack of order, usually by building long elevated expressways.

The action of creating a conceptualized place would, like all other actions of a city, be the result of negotiations, and generate negotiations of its own. These negotiations would tend to become public only in their final stages, when they are close to influencing the conceptualized place of the city and hence affect urban policy. As this conceptualized place would determine the rules that the city is expected to live by, there would be considerable attention paid to all that influences it. This would include interest groups that build campaigns for changes in rules, and thereby implicitly the conceptualized place underlying those rules. A group may want rules that exclude some others from using public places, such as a high entry fee for public parks. The conceptualized place of a city that generates such rules would allow for spatial differentiation to consolidate economic differentiation. The media could play a significant role in this negotiation, by presenting the cases made out by various groups. At times the media could itself become an interest group influencing the conceptualized place that underlies how a city sees itself and what it does about it.

The urban and the city

The analysis of places that go to make a city helps form a more precise distinction between the urban and the city. The urban is the entire set of processes of agglomeration and polarization and their consequences. The actions of the urban process take place over a wide range of places, from the villages where labour is mobilized, to the cities that are the destination of that agglomeration, and, in the case of global circuits, even further to other cities beyond national boundaries. The city is the contiguous collection of places, where individuals and groups are brought together by a substantial process, or processes, of agglomeration. The actions of the city, in contrast to those of the urban, are largely confined to this collection of contiguous places. Thus even as the urban thinks and acts global, the city may think global but its actions are predominantly local.

The conceptualization of the urban as global and the city as local defines the interpretation of the effects of globalization on the city. The processes of agglomeration and polarization and their consequences, which constitute the urban, are deeply influenced by the circuits of globalization. Those who control of the urban process may need to consider interventions at several points in the global process, whether it is labour conditions at the point of manufacture, or the demands of fashion at the point of final sale. Cities that are the destinations of some of these processes of agglomeration would also have to take the demands of the larger circuits on board, but these demands would be evaluated in the context of local interests. Cities may, and often do, want to attract global capital, but they would consider its effects, both beneficial and otherwise, on the various local places that constitute the city. Their response to this assessment would also be largely limited to local actions, such as the grant of land at concessional rates.

The focus on the local is also ensured by the collection of contiguous places that make a city. Agglomerations in the urban process can change the character of existing places in a city or simply add new ones, but there could be contiguous places that have been absorbed into the city which may not easily find a place in the processes of agglomeration. This phenomenon is omnipresent in cities of the global South that have grown horizontally to absorb villages within them. A few skilled artisans may be able, for extended periods of time, to continue operating in their essentially rural ways. Even for others who are forced to immediately deal with the processes of agglomeration, their rurality may extend into whatever role they find in the process of agglomeration. When the land of the village becomes a part of the city, there is no further scope for agriculture as the village knew it, and agriculturists lose their main occupation. In most villages that were absorbed into Indian cities, agriculturists were forced to rely on the few rural occupations that could be continued in the city, such as dairying. The existence of such rurality within the city extends beyond the economic domain. The village that is absorbed into the city could find ways to retain, and assert, its identity. This could be done through the celebration of village festivals in the heart of a city. The contiguity of the rural with the urban within a city ensures that the city cannot be identified with the urban alone.

The focus on the local is also seen in memories of a city. Some of the memories of the city could go back to times before it became the destination of major agglomerations, and have rural elements in it. These memories may be most evident in parts of the cities of the global South which were primarily villages not very long ago. But memories can go back a long way, with even large cities of the global North identifying some of their places with their rural past. The places, like Greenwich village in Manhattan, may have little in them that emphasises their rural past but their names could still reflect that memory. The memories of a city would include those associated with previous agglomerations. Cities with a colonial past typically retain memories of that period and may even prioritize that past to see the present as post-colonial.

The role of memory in defining places of a city also extends the life of a city. It is possible that the agglomerations that cause the rapid growth of a city may themselves not survive. Centres of colonial government attracted employees to run the colonised regions, and they, in turn, drew workers to build and maintain that infrastructure. Many of these processes of agglomeration died with colonialism, but the structures they created did not necessarily do so. Several cities in South Asia retain the imprints of colonial influence, including cantonments that now serve new masters. Thus while agglomerations of the urban process do die, cities do not necessarily die with them. Indeed, memories of cities can keep them alive long after their agglomerations have ceased to exist. This can be seen in ancient cities. The agglomerations that made Nalanda, in the east of India, a knowledge city from the 5th century CE to the 12th century CE, have long been dead, but it is still recognized, if only by its ruins, as an ancient city.

The conceptualized place of knowledge that Nalanda is now identified with need not have been the only place in that city in its prime. Nalanda would have been, like other cities, a collection of places, each of which had its own identity. Even as cities are identified with a conceptualized place, its other places and their identities do not disappear, and may even be asserted from time to time. The first two decades of the twenty-first century have seen a part of the new elite in Bengaluru trying to present the conceptualized place of the city as one centred around the information technology industry. This did not stop other places in the city from asserting their identity. This may have been most visible in the festivals of villages absorbed into the city, but other places, such as those occupied primarily by specific ethnic groups, can also assert their identity, usually through their festivals. In the midst of these multiple identities, it is possible that the city would be remembered by the identity associated with its conceptualized place, but this memory would itself be mediated by what the future wants to remember the city as.

The identities of the various places of a city are not immune to contestation. The scope for such contestations can increase with the degree of polarization, which emerges from a variety of processes. There can be the polarization between the new agglomerations seeking places in the city and the agglomerations that created older places in the city. There is the polarization between different players, say, blue collared workers and company executives, within processes of agglomeration, each seeking their own place. There is also the polarization between those who remember being a part of an agglomeration and those in villages within cities who do not have any such memories, and believe they have always been in that place. The actions of individuals and groups in the collection of places that make a city need not always be consistent with each other, calling for negotiations in different spaces to ensure the city does not slip into incompatible actions and conflict.

Wimoa, having come to the city as a part of a process of agglomeration, would very much be a part of the urban process, and the negotiations in the city that emerge from it. But she is more likely to be at the receiving end of these negotiations rather than being able to influence them. She is very unlikely to know much about the negotiations that go to make the rules for the city. Yet she may not be able to ignore the more local negotiations around whether the rules are to be followed or not. Should she pay the bribe that is being demanded of her family for not following a rule she had no idea about, or should she fall back on her identity group to protect her? Should she join those challenging members of another ethnic group threatening people in her neighbourhood, or should she try to slip away? Everyday life for Wimoa would include negotiating her way through the conflicts between places in the city, and between those who live in these places and the authorities who are expected to enforce the rules derived from the conceptualized place. For Wimoa, these negotiations – formal and informal, minor, and major – come with the suspense over what they will do to her, a suspense that can easily slip into fear.

References

Ackroyd, Peter, 2001. *London: The Biography*. London: Vintage.

Agnew, John A., 2011. Space and Place. In: *The Sage Handbook of Geographical Knowledge*, by John A. Agnew and David N. Livingstone, (eds), pp. 316–330. London: Sage.

Halbwachs, Maurice, 1950, 1997. *La Memoire Collective*. Paris: Albin Michel.

James, Lawrence, 2013. *Churchill and Empire: Portrait of an Imperialist*. London: Weidenfeld & Nicolson.

Massey, Doreen, 1994. *Space, Place and Gender*. Cambridge, UK: Polity Press.

Morris, Morris D., 1983. The Growth of Large-Scale Industry to 1947. In: *The Cambridge Economic History of India*, by Dharma Kumar and Meghnad Desai, (eds.), pp. 551–676. Cambridge: Cambridge University Press.

Sinha, A, & J. Singh, 2011. Jamshedpur: Planning an Ideal Steel City in India. *Journal of Planning History* 10(4), pp. 263–281.

Sudjic, Deyan, 2005. *The Edifice Complex: How the Rich and the Powerful – and Their Architects – Shape the World*. New York: Penguin Books.

Tønnevold, Camilla, 2009. The Internet in the Paris Riots of 2005. *Javnost – The Public* 16(1), pp. 87–100.

6 The happenings of urban suspense

A view of the urban and the city through the lens of action will find a prominent place for personal experience. A person's urge to act prompts an action followed by the experience of carrying it out in a specific situation. Even when the action is intended, it has to contend with a variety of circumstances that could lead to unintended consequences. At the very outset a person's urge to act may come up against the urge of others to act in a contrary direction. As each person works out the power of her conviction vis a vis that of the others, the options she has, and her sense of fairness regarding that action, there would, in effect, be a negotiation even if the individuals have not met. These negotiations could take place through several media, from the market to social pressures to gossip. The negotiations would continue through the entire process of carrying out the action and managing the consequences. Each stage in this process would be marked by considerable uncertainty.

As Wimoa steps out from her home in the city to get a job, she may well be burdened by a considerable degree of self-doubt. Does she really need to step out to get a job so soon after coming into city she does not know, or should her husband have given her more time to come to terms with her new surroundings? She may have convinced herself that she does want a job, or in any case needs it, which would bring the uncertainty of where to try for one. Prompted by her friends, and forced by her family, she may focus on a particular factory gate where she could try her luck. This would be followed by the uncertainties over the availability of jobs at that factory, the number and intensity of others trying for similar jobs, the persona she should present at the gate, the attitudes of the person hiring, and much else. The uncertainties are not confined to potentially life-changing actions like getting a job in a new city. Simple everyday actions like finding her way in the city would have their own elements of uncertainty, beginning with Wimoa's sense of direction. To the uncertainties of intended actions, must be added those of what happens to happen. The unpredictable happenings of everyday life in a city can be quite considerable, especially for someone like Wimoa in a city of the global South. Apart from the uncertainties of her job, if and when she gets one, her everyday life could be disrupted by a variety of other events. Heavy rainfall could flood her house, bands of violent youth could get unruly and disturb her neighbourhood, she could be the target of harassment on the street, and the victim of other events with varying degrees of predictability.

DOI: 10.4324/9781003196792-6

From the point of view of an objective outsider – a persona that analysts of cities sometimes like to take – it is important not to get bogged down in these uncertainties, and to focus on what finally happens. The city is then analysed in terms of all that has happened rather than what could have been. Those carrying out the actions that go to make a city do not have this luxury. Each of their actions would have varying degrees of doubt built into it. Even the most commonplace actions like taking a bus to work would have an element of the unpredictable to it, ranging from taking the wrong bus to the possibility of an accident. When the city is seen as a collection of actions – and not just what finally happens – these uncertainties cannot be brushed aside. If nothing else, the suspense over what could happen would influence the urge to act. The mere possibility of something untoward happening could stop Wimoa from wanting to enter some parts of her city alone after dark. This suspense is so widely accepted a part of everyday life in the city, that it is easy to be taken for granted. Yet, a complete picture of a city would include what it does to the suspense its citizens experience when carrying out the actions of their everyday life. The role of this suspense in the city is multidimensional and substantial enough to require more attention than passing references, such as recording some cities as being safer than others.

Urban suspense

The suspense over an action is, at one level, linked to the degree of uncertainty that goes with it. The uncertainty can be the result of conditions in the city as a whole. In a war-torn country there can be considerable suspense around the uncertainty on whether a city will be bombed or not. Yet it is not uncertainty alone that generates suspense. In calmer, more routine, times, Wimoa could feel a degree of anticipation and suspense about the arrival to her city of a cousin she is fond of, though she knows all the details about the time and point of arrival of the bus her relative is taking to the city. Experts go a step further and speak of the paradox of suspense, where a person can feel a degree of suspense despite knowing the outcome (Yanal, 1996). There is, in fact, a degree of suspense over any anticipated event. In what is sometimes considered the standard account, Ortony, Clore and Collins theorize that suspense is composed of hope, fear and a sense of uncertainty (1990). Adapting their theory to Wimoa's urban situation we can understand her suspense as being composed of hope (the pleasurable prospect of an action in the city affecting her in a desirable way), fear (over the prospect of an action in the city affecting her in a way that is undesirable), and the uncertainty arising from both not knowing which of the two actions would happen as well as the possibility of a completely unexpected happening drawing her away. Wimoa would hope her cousin reaches safely, fear that she may not, and not know which of the two would finally happen or whether an unexpected assignment at work would keep her from meeting her cousin.

The suspense that permeates every expectation, big or small, in a person's experience is far from being immune to urban processes. The processes of agglomeration and polarization generate expectations from those who move to the city. There are hopes built around what they expect the city to offer them, fears about those expectations not coming through, and a great deal of uncertainty about what would actually happen. There are also likely to be fears about aspects of the city they know nothing about, a part of a larger fear of the unknown. They may hope that they will be accepted by those who are already in the city, but fear that they may not. What is more, the consequences of agglomeration and polarization would generate their own sets of hopes, fears and uncertainties. The rapid growth of a city would require substantial numbers of people joining its processes of agglomeration. Yet the resultant influx of workers from elsewhere could cause older residents to fear that the new entrants into the city would get the available jobs.

These points of suspense could emerge in any of the spaces of the city. The popular narrative provides a great deal of emphasis to the suspense in absolute space. Will a person face any difficulty in getting her own place to stay in the city? If she has to rent one to begin with, when will she be able to own her home? Each of these expectations is built around hope but also has its concomitant fears. If she buys a house, can she be sure the title is genuine? The suspense over these issues is usually sought to be reduced to its economic dimension. The focus is on the material goods a person can afford, and the set of material goods she will choose. There are also economic tools designed to address some of the fears that go with the buying of a product. An effective system of guarantees is expected to overcome the fears of the failure of a product. Brands are also expected to reassure the customer about the quality of the product.

The suspense of material space is, however, not an economic matter alone. Arrangements for a place to live in the city can be built around elements that are not entirely economic. There are empirical studies of the entire process of migration from the village. One such process – one that Wimoa would identify with – would begin with the husband seeking employment in the city (Pani & Singh, 2012). The immediate need would be for a place to stay in the city while searching for a job. For those men who cannot afford accommodation in the city there is a need for alternative arrangements usually based on social networks. He could live with a relative, a member of his caste, or simply someone from his village. This sharing of a small tenement could continue through his first period of work in the city. This places considerable pressure on the woman in the urban household. She has to cook for the entire household, including the new entrant, typically in addition to doing a factory job herself. Since patriarchy would limit, if not rule out, men helping with household chores, the next best option for the new entrant to the household would be to get his wife to the city to help out. Once the wife comes in, there would be a need to contribute more to the economics of the household. This would exert pressure on the wife to find work. In this entire process, of cramped living and long working hours, they would not be able to find a safe place in the city to leave their children when they are away at work. They would then prefer to leave the children in the village with their grandparents until they are old enough to take care of at least their basic safety in the city. Once the children are ready to come to the city the husband and wife would ideally move to another tenement. Over time the new home could become a place for other migrants following up on their hopes in the city.

This process has an entire set of hopes, fears, and uncertainties that the migrating family has to deal with. At each step of this process there is the uncertainty of whether others in the social network would act in a way that would enable the family to migrate. This would include both the part of the network that resides in the city as well as the relatives in the village who are to manage the children until they are old enough. This would be in addition to the usual uncertainties of getting a job in the city, getting used to an urban way of life, and gaining confidence over the safety of their children in an urban setting. These uncertainties throw up myriad fears and the need to overcome them in order to realise the hopes the city offers.

The uncertainties, hopes, and fears would themselves vary a great deal. There would be those who are even more economically vulnerable than the persons Wimoa could identify with. They could be illiterate, unskilled and not earning enough to bring their families to live in the city with its high costs of living. They cannot hope to complete the process of migration even in the splintered form that could be done by those at Wimoa's economic levels. These extremely vulnerable workers would keep their families in their villages even as they work in cities. They would seek to earn the relatively higher wages of the city and spend in the relatively lower cost economy of the village. They could live crowded in with other similarly placed men in a shed at, say, the construction site at which they work. They would typically limit their expenditure to a bare minimum in order to enable them to send most of their earnings back to their families in the village. The only indulgence they would allow themselves would be phones that would allow them to stay in touch with their families. The use of the mobile phone too could vary depending on the economic condition of the worker in the city. The most vulnerable could use innovations like a missed call to message their family in the village. There could be a call made as soon as the worker reaches the city to signal the family that he has reached safely. Since the family members would not pick up the phone there would be no charge but the message would have been effectively conveyed. As the workers become less vulnerable they may be able to invest in a smart phone which would allow them to see their children when they talk to them.

Being among the most vulnerable in the city these workers have to deal with a wide range of fears. There would be the fears of not being employed for a sufficient number of days to make their trip to the city worthwhile. If they belonged to ethnic groups other than those that dominate the city – as would tend to happen if they had moved long distances from their village in search of jobs – they would have their own fears. They could fear being attacked by local ethnic groups. And they would be vulnerable enough to be targeted on a variety of, frequently false, charges ranging from being perpetrators of crime to taking away local jobs. These fears could be overcome by the hope that the city would provide them a better life. The components of the better life the city provides need not be in that urban territory. As the workers hope to spend their earnings in the village, their ambitions could be tied to the village. They may hope to use their earnings from the city to improve their status or that of their family in the village. A study has illustrated the case of migrant workers from a village in Bihar in eastern India pooling in their resources to get one of their own elected to a local office in their village (Haque, 2018).

The process of gaining a small foothold in the politics of that village by the former marginalized workers was by no means easy. The migrants made an effort to use their earnings in the city to buy more residential land in the village to build larger houses for their families. Those already in power in the village saw the size of the houses of the migrants as not just a matter of convenience but also a statement in representational space; a statement of their new status within the village. Those already in power got together and collectively prevented anyone from selling land to the migrants. The migrants responded to this challenge to their ability to use their houses to represent their change in status by building multi-storeyed houses on the small residential plots of land they already owned in their village. As the battle moved into conceptualized space the hope was one of greater status and power, and the fear was that of the loss of it. The uncertainties about the outcome can extend well beyond these spaces, given the tendency to violence to settle such conflicts of over issues of status in some rural contexts.

The tendency to build larger houses as a statement in representational space is by no means confined to the lower end of the economic hierarchy. Among the more striking of these examples is that of a building in south Mumbai, which is home to one who frequents the Forbes list of richest individuals in the world. Popularly estimated to cost over one billion dollars it is, at least in local discourse, believed to be the most expensive bit of real estate after Buckingham Palace. Home to a six-member family, its upkeep is said to require 600 workers. Even so emphatic a symbol of economic power in the conceptualized space is not devoid of its elements of suspense. Taxi drivers in Mumbai have developed a narrative that sees the helipad on top of the 17 storeyed building as the result of a fear that getting from that building to a hospital in the event of a medical emergency would be difficult in Mumbai's traffic.

The validity of the taxi drivers' narrative is clearly impossible to confirm. It is difficult to see anyone who the multibillionaire may have confided in, confiding in taxi drivers. But the narrative points to the widespread recognition of the overwhelming significance of the neighbourhood to the suspense individuals face, including the most powerful and exclusive persons. The element of relational space is much more evident in the lives of those who do not have the means to isolate themselves from their neighbourhoods even when they would most like to do so. The suspense of passing through a violent neighbourhood, let alone living in one, is dominated by fear. Proximity to a violent neighbourhood brings with it a threat to lives, the possibility of younger members of the family being drawn into drugs, and becoming either perpetrators or victims of violence or both – all very real fears in the more difficult neighbourhoods of a city. The hope that goes with this fear is often no more than one that none of the feared outcomes would come to pass. At the other end of this spectrum of relational space would be the hope that the neighbourhood is free of crime. Further up the hope scale, the neighbourhood can be expected to provide other benefits such as good schools.

The suspense of urban life is, arguably, most varied in the organic space of the city. At a somewhat benign level a person's suspense may be over no more than the hope of finding exotic food she would not find elsewhere, which would necessarily be accompanied by the fear of being disappointed. The suspense would be greater if she were to fear the vaccination she has given her child has past its expiry date, leaving her with the only option of hoping the doctor has not made a mistake. It could move on to the uncertainty of bodily harm through a traffic accident. The fear of a sudden and unexpected loss of life could be heightened if she were to live in a city which is near a war zone, or is simply known to be a target of terrorists. Here again she may not have any option other than to hope that those tasked with protecting the city have done their job well.

The elements of suspense in the different spaces of life in the city are deeply intermingled, sometimes reinforcing one another, and at other times offsetting each other. The suspense over a neighbourhood in the relational space would be accentuated if the person did not have the option, in the absolute space, of finding a home elsewhere. There are few greater challenges than trying to bring up a family in a violent neighbourhood, knowing all the while that there is no other option. In other cases, the hope of earning enough in the city to buy a home, whether in the city itself or back in the village, would create its own suspense accompanied by the fear of failure. But as long as this suspense is dominated by the hope of having her own home, it could help offset the fact of, say, having to live in a violent neighbourhood. In this intermingling of suspense, time could play a significant role. The very nature of a place could change over time. This could happen even within the course of a day with some areas of a city being more unsafe late in the night than they are in broad daylight. This could bring its own suspense. A woman could ask herself at dusk whether it is too late to be out on the street alone, and hope that it is not.

There is a fairly widespread, if sometimes unstated, belief that a city should necessarily reduce urban suspense; there should be an effort to reduce the levels of uncertainty. This leads to a variety of measures in that direction, ranging from weather forecasts to more predictable processes of gaining employment. Yet there are aspects of a city that build on the maintenance, if not accentuation, of suspense. The anticipation of a sporting event or the performance of a rock band brings with it suspense over the hope of a ticket, the hope of enjoying the performance, or of simply being there. Those working for the success of the event would actively contribute towards increasing the anticipation and the suspense around it. Less pleasurably, there could be efforts to increase the suspense around the fear of an unwanted event. Terrorism is frequently directed at increasing the fear of unexpected violence and death. The objectives of increasing the suspense of those seeking to promote a particular sport, or other entertainment events in a city, are built around hopes and hence vastly different from the promotion of fear in the suspense that terrorists seek. Yet in the pursuit of these sharply contradictory urban goals, the two may, in fact, traverse the same spaces of a city, in the process sometimes feeding into each other.

Sport and the suspense around hope

The suspense around hope can have a pleasing dimension to it. There is the anticipation of a favourable outcome, and the joy if it does come about. The difficulty in real life negotiations around hope is the possibility of failure, especially if the failure has serious consequences. There is then much pleasure to be had from a construct that generates the anticipation and other pleasures of hope in negotiation without the possibility of serious adverse consequences. Sports usually serve this process. A sport lays out a common set of rules and neutral monitoring to ensure that there is a sense of fairness to the game. It defines the boundaries of the playing field in a way that ensures that the options are not unlimited. This allows for the outcome to depend, ideally, entirely on the ability of the adversaries. The consequences of the failure are also kept within limits, though the limits would depend on the norms of the city where the sport is played. The cost of failure in a gladiatorial contest in ancient Rome was death for the contestant. In modern cities the costs of failure are fortunately nowhere near as extreme, though it may not always be possible to believe "it's just a game". And the entire process of controlled suspense is carried out in a fixed timeframe that would determine whether hope is realized, (usually in a victory at the game), hope is lost (usually in the loss of the game), or hope is put aside (usually when the game is drawn). And there is a closure to the suspense when the game is over and the results are known.

Individuals in the midst of the diverse, and sometimes intense, negotiations of the city can see in the controlled suspense around sport a reflection of real-life negotiations. In an individual sport they can identify with one or the other of the players and take the victory or defeat of their player to be their own. They can get the pleasure of suspense around hope, with limited consequences of failure. This identification can extend to larger identities that individuals associate themselves with, including nations. Sporting events like the Olympics or a soccer World Cup can generate national pride in vast multitudes of people around the world. As cities have grown in importance, so has the identification with them in sporting events. Football leagues in England have a prominent place for teams associated with cities, or even parts of cities. In countries where commercially driven metropolises have been a relatively later phenomenon, team sports based on cities have also emerged later, as in the case of the Indian

Premier League in cricket. The association of the team with the city is primarily in the conceptualized space, with players being drawn from all over the world, and shifting from one city team to another, often based on the best price that is offered to him or her.

The course of a sport is quite significantly influenced by the demands of cities. This can be seen in all three routes to experience the controlled suspense of a sport: as a player, as a spectator around the playing area, and as remote spectators through media. The experience of a player changes with the nature of the agglomeration that drives the places of a city. In a less commercially active city, the individual sportsperson would not be able to generate the resources from sport to make it her profession. She may have to make do with sport being a secondary activity. In cities that are driven by more competitive market-led agglomerations there would be economic benefit in associating a product with the attractiveness of the controlled suspense of sport. The most common form of doing so is to get a successful sportsperson to endorse the product. The association with the sportsperson and her sport would take the product into a variety of spaces in the city. The endorsement could take the product into the imaginations of the lived space, where the sportsperson is an icon. The influence would also extend to the organic space of a city, as the sportsperson endorses what people should eat and what they should be seen to be wearing.

The process of endorsement is based on the link between the player and the spectator. It is expected that the fan of a player would extend that support to the products he or she endorses. Beyond their loyalty to individual players, the active involvement of spectators can also transform the experience of the sport. The passion of this involvement can contribute to the spectators at the stadium, where the sport is being played, being seen as a part of the game. Their cheering can push their favourite players to improve their performance, and even if that does not happen, it would contribute to the atmosphere around the sporting event. The integration of the spectator in the stadium with the sport has been enhanced by the third form of experiencing the controlled suspense of a sport, that of the remote spectator. The visual media has made the spectator at the stadium a part of the sporting event. The response of these spectators can be used to enhance the suspense of the sport. In the process the remote spectator becomes the one who can see the game in its entirety, not just the actions of the players but also those of the spectators at the stadium.

The centrality of the remote spectator strengthens the influence on the sport of the processes of agglomeration that drive the places of a city. The time that a city provides a spectator to spend on a sport is affected by the nature of the underlying agglomeration. Agglomerations that are state-led and less dependent on market competition tend to be able to give their workers greater time to be remote spectators at their favourite sporting events. Agglomerations that are driven by aggressive economic competition, in contrast, typically claim a greater share of a worker's time. The increased competition of a globalized city may have, in fact, led to spectators finding less time for the controlled suspense of sport. These pressures on the time available to the remote spectator, who is now central to the sport, has affected the nature of the game. It has contributed to a premium on capturing the entire excitement associated with a sport in as short a time as possible, without losing out on the development of suspense around the hope of victory. This has led to an increase in the popularity of shorter games. In response, the organizers of sport have been willing to alter the rules of the game to suit the limits on time available to spectators in cities driven by competitive agglomerations. This is perhaps best seen in the case of cricket. As a sport developed in the colonial era, its games were conducted at a leisurely pace, running into several days, leading to the modern form of Test cricket spread over five days. The difficulty in maintaining interest in such a sport in cities with great pressures on the time of remote spectators led to the creation of shorter formats of the game. One-day cricket, with its own set of rules, including 50 overs a side, emerged in the 1970s. This was followed by the even shorter 20-overs-a-side cricket. The shorter versions of the game are usually played under lights so as not to come into conflict with the working hours of the city.

The controlled suspense of sport is not infallible. It is possible for the close identification of different groups in a city with their teams to result in the conflicts of a city being taken into the sporting event. The suspense then becomes uncontrolled and can have wider consequences for the relations between these groups in the city. For decades before Indian independence, cricket in Bombay (now Mumbai) was played between teams like the Hindus, the Parsis and the Muslims. This reflected the communal divisions in the city at that time. And in the communally charged atmosphere of the decades before Indian independence, a cricket match had the potential to turn into a spark that lit much larger fires. Gandhi, as a part of his larger effort to reduce communal divisions that had taken a toll in violent riots, campaigned against the tournament that pitted one community against the other on the cricket field (Majumdar, 2002). Even in the twenty-first century it is not unknown for cricket matches between India and Pakistan to lead to the suspense generating uncontrolled consequences. As sports gain a prominent place in the lived space of a city, the task of controlling its suspense can become more difficult, sometimes leaving a longer-term imprint on the city.

Beyond the scars of conflict sparked by the uncontrolled suspense of some sporting events, sport does have a more positive longer-term impact on the city. The infrastructure created in the absolute space of a city can influence the way a city is seen, as in the experience of tennis and Wimbledon. In the case of major sporting events that move from city to city, like the Olympics, the infrastructure that is created for the event can typically be used to meet other ends after the event is over. The choice of a city to host a particular event also adds to the civic pride of its citizens. This pride can contribute to an improvement in the quality of civic life even after the sporting event is over. Preparing a city for the global attention a major sporting event brings may also result in it addressing situations related to organic space that may not otherwise have been on top of the mind of local planners. Beijing did take several steps to reduce air pollution before it hosted the Olympics.

Apart from these material benefits, the controlled suspense around hope helps citizens cope with the pressures of the never-ending negotiations of a city. It can help keep the potential for conflict between divisions within a city to the boundaries of a sporting event, even as a failure to control the consequences of that suspense can deepen the divisions.

Terrorism and the suspense around fear

The suspense in a city built around fear can take a variety of forms. In a racially, or otherwise, divided city the fear of the other could be a major source of concern. The fear could take on a more individualistic character as in the fear of being molested in a dark street. And there are a variety of interests that can be served by stoking these fears and thereby enhancing the extent of this variety of urban suspense. Politicians who are seeking to mobilize one section of a divided city often do not hesitate to step up the fear of the other. The fear of a rise in the crime rate is also a subject of what is usually considered legitimate political activity. In terms of the stoking of everyday fears these common, usually political, actions are probably the most widespread. Once we add the element of uncertainty to the equation, terrorism, arguably, comes to the forefront of the urban discourse of fear. The fear of terrorism goes well beyond what would be predicted by the probability of being the victim of a terrorist attack. In most cities the probability of being a victim of terrorism would be well below the risk of being the victim of a traffic accident. But there are aspects of the fear of terrorism that cannot be reduced to the cold rationality of probabilistic analysis. A person's fear of being the victim of a terrorist attack is not simply the result of the magnitude of the destruction caused by that act of terrorism. The fear is also enhanced by the heightened suspense over when and where such an attack could occur. The uncertainty around the event, the fact that a terrorist attack can in the mind of the potential victim happen anywhere at any point of time, adds to the level of suspense. It suits the terrorists to further enhance this fear of the unknown through the murderously cynical manipulation of the spaces of a city.

The specific instruments terrorist groups can use to carry out their destruction are known to vary a great deal. They can range from suicide bombers, to flying a hijacked aircraft into a commercial or military centre to driving a vehicle into a crowded pavement. Stepping back from the immediate instruments, terrorist actions are typically built around at least three other considerations: they target the fear of the unknown, they link this fear to a well-defined other, and they seek to use violence in a way that allows clusters of individuals to target much larger and more powerful groups. These elements are not mutually exclusive. It is quite possible for a terrorist action to simultaneously represent an extreme ideology of hate of the other, and operate through the psychological domain to target larger groups. These elements of a terrorist attack make cities the obvious targets. The large number of people in a city ensures a degree of anonymity, with little being known about each other. As Wimoa enters the city she will come into contact with persons from ethnic backgrounds she has never interacted with before, in fact ethnic backgrounds she may not even have known existed. As she manoeuvres her way through this vast mass of people she knows little or nothing about, it is not difficult for her to slip into using stereotypes. Social and political groups in the city can manipulate these stereotypes in ways that enhance Wimoa's fear of the unknown. In an atmosphere of enhanced fear, a single act of violence can generate fear of a magnitude that shakes up the city and areas beyond it. It suits terrorists to enhance this fear. Whether they are able to do so depends on the extent to which their acts leave an impact on the multiple spaces of the city.

In the conceptualized space, a terrorist act is designed by its perpetrators to meet at least two objectives: to strike terror in the enemy and to attract more recruits to the murderous cause. Terror is a means of getting the attention of major powers by a less powerful and smaller group. The magnitude of the terror is enhanced when instruments of everyday life are weaponised as in the use of commercial aircraft to bring down the World Trade Centre towers. The process of generating terror is simultaneously designed to be attractive to potential young recruits. The terrorist actions can be presented as acts of bravery and sacrifice, when there is a loss of life of the perpetrators. This sacrifice is associated with a larger cause through a more elaborate exposition of ideology and victimhood, usually both.

The portrayal of much of a larger ideology into a specific terrorist act depends, to no small extent, on the choice of target. The terrorist would prefer a target that would be immediately associated with all that their enemy stands for. The choice of target in conceptualized space – the space where symbols of specific social trends emerge – often provides a clear indication of the stated ideology of the terrorist and the actual goal. Twenty-first century terrorism is often interpreted as a fight between religions. For this message to be conveyed through a terrorist act the target would have to be major religious institutions. If instead the object of the terrorist action was the dominance of major countries and groups the target would be symbols of the economic and military power of those countries or groups. The focus on the World Trade Centre and the Pentagon in the 9/11 terrorist attacks or the terrorist attack on Indian parliament would all suggest the primacy of political and economic power over religious conflict in twenty-first century terrorism. The focus on the political and the economic ensures terrorists focus on large urban metropolitan centres that house the institutions that control these activities, or are symbols of their success.

Within these metropolitan centres the terrorist attacks would aim to enhance the fear of sudden and unexpected death. One of the responses to a sudden attack would be to escape from the site of the attack. The scope for such an escape would be limited if the organic space around a potential victim is limited. She would not be able to see a way out of the site of the attack. Terrorists can choose to enhance the fear of an attack by focusing on sites, such as passenger aircraft, where the organic space is limited to what a passenger can see within the cabin. The fact that once a plane is in the air she has no option of stepping away makes her sensitive to even the slightest chance of a disruption in the normal activity on the plane. A terrorist attack in an aircraft not only causes immediate destruction but also spreads the fear to other passengers in other aircraft. The use of organic space is also not restricted to areas which are confined by physical boundaries. The ability to get out of a site can also be severely restricted by the suddenness of an attack. A sudden attack could take a victim by surprise in the midst of a large crowd. The limited time to get away confines the victim to a small space and in the process generates a fear of even routine activities. The spate of killings by driving a vehicle into a crowd on a pavement is designed to generate a fear of everyday places.

The extent of fear a terrorist attack can generate is also the result of the association of a space with a terrorist action. The sites of a terrorist attack are usually associated with terrorism for long after the attack. This memory and association is, ironically enough, strengthened by the very humane need to build memorials of the victims of the attack. The association of a space with a terrorist attack is also influenced by extending the geographical spread of those who are directly affected by the attack. By targeting events in a city that attract crowds from not just all parts of the city but also those in other places, terrorists can ensure that the friends and relatives of the victims are spread out across a much larger area than a single city. Terrorists can then tap the relational space to target events in places that would be associated with the act for a long time, as well as have a much broader spread of fear after the attack.

The attack itself is in an absolute space with clearly defined boundaries. The direct impact in that space is determined by the loss of life that it causes, in addition to the destruction of the physical space and all that it conveys. There is also a second absolute space that the terrorists are interested in. This is the absolute space in which those who are targeted receive the information about the attack and feel the fear. This space, say in front of a television at home, is occupied at some time and empty at other times. It is in the interest of the terrorists to carry out the attack at a time when it when this space would be occupied. The launch of the Mumbai terrorist attacks on November 26, 2008 was timed to catch prime time television.

Escapism

The heightened suspense built around hope, fear and uncertainty necessitates a response from those in the city. The suspense around hope, particularly when it is within a fixed timeframe as in a sport, would be seen by many as desirable. Wimoa, even if she does not fully comprehend the nuances of a sport, may be happy to join in with celebrations of a victory. In contrast, the suspense built around fear is by its very nature not desirable. A person would then ideally like to enjoy the suspense of hope as long as it does not get to be a strain even as she acts to reduce the suspense built around fear. This response to the varied and interrelated suspense of living in a city would be influenced by individual traits. There would be those who would be expected to be confident enough of realizing their hopes that they would dismiss their fears, even as some others are presumed to be more engulfed in their fears to a point of being afraid to hope. There would be some who would usually try to reduce uncertainty through the use of rational analysis, while some others may do no more than fall back on prayer. It is important to remember that there can be considerable variation in individual behaviour across place and time. On a vacation by the seafront, a person may allow herself to act on an impulse, something she would not normally do in a formal discussion in her office. As with much else in the city, the response to suspense is best explored in terms of actions rather than individuals; allowing for the fact that individuals can react in varied ways to suspense across different situations.

If we were to review the phenomena that we commonly see around us in urban life but do not find a corresponding place in analyses of the city, escapism would rank quite high on that list. Escapism is usually treated as no more than an avoidable personal trait. Even the few who think it is worthy of analysis see it purely as a negative phenomenon, that prevents acting in a way that addresses the reality of specific situations. To the extent that it takes the mind away from reality, it is not expected to be of much use in analysing specific situations. Most among the relatively few researchers in this domain would go along with Vorderer that "In its core, escapism means that most people have, due to unsatisfying life circumstances, again and again cause to 'leave' the reality in which they live in a cognitive and emotional way" (Vorderer, 1996, 311), cited in (Henning & Vorderer 2001). But escapism need not be a response only to undesirable situations. As an action it can be a part of situations that are, on the whole, very positive. It is quite reasonable when expecting the result of, say, an examination, to ease the pressure by thinking of, or doing, something else. In such cases escapism is not merely a response to unsatisfying life circumstances but to the hope of doing well in the examination, the fear of failing in it, and the uncertainty about what would turn out to be the result of the examination. Escapism would then be a response to suspense. The fear of unsatisfying circumstances would only be one of the phenomena that can prompt escapism; it could also be prompted by the uncertainty about a forthcoming event, including a much anticipated future event. We could then define escapism as responding to a particular point of suspense by thinking of, and/or doing, something else.

When seen in these broad terms, escapism can be an individual's response to her experience of suspense in several spaces of the city. In the absolute space a person can avoid a particular suspense by simply choosing to be somewhere else. There could be situations where this choice of escapism would have distinctly negative consequences. In order to find work in the city Wimoa would need to meet the agent at the factory gate and make her case for a job. If she has never done it before she may fear the agent, the possibility of failure, or indeed the entire process. This may prompt her to escapism by insisting someone in her family is sick and she needs to be with that person. Since this would rule out the possibility of getting a job, the consequences of this act of would quite undesirable. There could be other situations where escapism in absolute space – avoiding a presence in one place by being in another – can be quite positive. If a riot breaks out in a particular part of a city the instinctive response of most individuals would be to try to be somewhere else. Acting upon this escapist instinct would be a positive consequence for the individuals involved.

More common forms of urban escapism can be found in the organic space. A person could try to escape from the discomfort of being pushed around in public transport by listening to music through her earphones. She may use the same instrument to escape from the drudgery of walking from the metro station to her home. More sustained pressure may result in consistent recourse to acts of escapism. Some may choose to eat excessively as a way of getting their minds off issues of sustained urban suspense. It has been argued that binge eating is a form of escape from self-awareness (Heatherton & Baumeister, 1991). It has also been shown that laboratory stressors, as well as self-perceived stress, are associated with greater food intake (Groesz, et al., 2012). This phenomenon may have contributed to the crisis of obesity in cities of the global North, but it is by no means confined to those regions.

Escapism can extend to the lived space as well, when life in the city provides a prominent place for negotiating with the unknown. There are rational instruments that attempt to reduce the uncertainty of everyday life in the city. The forecast of meteorologists have considerably reduced the uncertainty around when weather would alter the everyday life of a city. But there are other urban uncertainties that are less predictable, or at any rate do not entirely rule out the possibility of an undesirable event. Data could tell us that the probability of a particular person being involved in a traffic accident would be very low. Yet in the mind of the individual in the city it need not be completely ruled out. This uncertainty frequently leads individuals to fall back on belief systems. These beliefs could be no more than irrational superstitions. They could range from, say, tying your right shoelace first, to new forms of witchcraft such as the ones that have been studied in urban settings in Africa (Mildnerová, 2016). This could result in visits to places that claim to represent the supernatural. More commonly, these places could claim to represent divinity.

It is also possible to resort to escapism by simply shifting thoughts and actions from one space to another. A person who faces several issues of suspense in the relational space at home could decide to focus her attention on the lived space of the city. She could spend more of her time entering debates on what the city should be, in the process becoming an activist to a civic cause. Someone who is faced with suspense in organic space, say an illness, could use escapism by contributing generously to a place of worship that is associated with a particular religion in the lived space. Wimoa could seek to escape the suspense in absolute space over when she and her husband will get a place of their own in the city, by falling back on organic space and spending time listening to religious music in a nearby place of worship. She may even go a step further and make an offering to a deity who, she has been told, grants such favours.

The existence of such non-rational belief systems in the city is often linked to the need to manage the unknown. These belief systems may begin as no more than individuals advocating non-rational practices to those facing issues of suspense. These practices, usually involving an offering, are carried out in the expectation of a specific hope to be realised or a possible feared event being averted. Some of these transactional practices can, over time, be extended to more formal institutions. Godmen in India have had considerable success with a variety of urban populations. And this is by no means confined to the less socially aware or the uneducated. There are Godmen who, with their own, sometimes quirky, justification of prescribed practices, have been known to have a considerable following among urban residents, including the elite (Mall, 2017–18). The existence of local non-rational belief systems can influence the practice of organized religion as well. There are places of worship which attract pilgrims of more than one religion. The dargah of Khwaja Moinuddin Chisti in the north western Indian city of Ajmer is a Muslim place of worship that has a considerable Hindu following as well (Burman, 1996). There are also more recent responses in cities of the global North to the need for places of worship for people with different faiths. Several multi-faith places of worship have emerged in the United States and the United Kingdom (Dalton, et al., 2006). This reflects the process of the practice of religions modifying their approach in response to urban change. It has been argued that religious organizations have to deal with the challenges of urban restructuring, religious restructuring and social transformation (Livezy, 2000).

Underlying the strengthening of old religious practices, and the emergence of new ones, is the possibility of a single response to multiple prompts to escapism. Each case of escapism may be in response to a very different suspense. One person may be seeking to escape the fear of an economic loss, while another may want to forget a personal emotional loss. But if they both go to the same place of worship, there is a single form of escapism that is catering to their need to escape from diverse sources of suspense. The agglomeration that grows a city brings together a multitude of sources of suspense, but the need to divert attention from them can be met by relatively fewer forms of escapism. This agglomeration of multiple sources of suspense into a demand for fewer forms of escapism provides one of the corner stones of the entertainment industry. Some of this escapism is explicit, as in the case of superhero movies centred around a character with no claims to reality. Most Hindi cinema is only a little less explicit, going in for unlikely, rather than impossible, characters. But even entertainment that claims to be realistic can be an instrument of escapism if it helps divert a person's mind from the particular suspense she is facing. Neo-realism in the cinema of Vittoria De Sica or Satyajit Ray delves deep into very real situations, but as long as the issues in the movie are not directly related to the suspense a particular person faces, it would still contribute to the escapism she is seeking.

The actions that provide routes of escapism influence the nature of a city. The competition between different routes of escapism that are on offer can have its impact on the structure of a city. The effort to promote theatre over film and other forms of entertainment could lead to the branding and development of specific places for it. In some cities the places associated with theatre may be no more than a complex with a sufficiently large auditorium, while in others it could be a street, as in Broadway. In other cities the branding of escapism can go beyond a particular movie, or series of movies, to the stars in them. This can influence the very appearance of the city. The citizens of the south Indian city of Chennai have been known to transfer their needs for escapism beyond cinema to particular movie stars, to the point where film stars have gone on to become chief ministers of the state. As a result the city has been known to be peppered with much larger than life cut-outs of film stars. The film and the associated industries (including the makers of the cut-outs) impact the economy of the city as well.

Oddly enough, escapism can use suspense as a means of avoiding suspense. This is done when situations of suspense are created within forms of escapism to attract an audience that is seeking to avoid another suspense. The element of suspense in films is a well-established means of attracting an escapist audience. Arguably, sport is better designed to meet this need. As we have seen, it carries the hopes of the supporters of two sides, the fear of failure, and the uncertainty about the outcome. Since it comes with a fixed timespan the suspense will also usually end at a predetermined time. Cities can also choose to be associated with specific teams in order to build their own brand equity. The identity of Manchester, particularly outside Britain, has been enhanced by its soccer teams. Over time these teams can build their own social identities around groups of fans. This can, in turn, affect the everyday life of a city as when large crowds gather for a match or, less desirably, when the behaviour of rival fans leads to hooliganism.

Incompatible actions

Attractive as escapism may be as an immediate response to suspense, it is rarely the sole response. The pursuit of escapism, in addition to being an action in itself, could prompt other actions. While standing in an overnight queue to buy a ticket for a popular performance, a person would interact with others waiting with her. She could help them, be helped by them, and get into conversations that help her pass the time. In the process she could discuss her hopes and decide on the actions that can be taken in pursuit of those hopes. She could go on to talk about her fear of not getting a ticket despite having waited the whole night, and perhaps other fears as well. Wimoa would in all likelihood not be able to afford the ticket to the performance and would not thus find herself in this queue, but she would have other hopes she seeks to pursue. She could hope to get a job and earn an income of her own. She may well belong to a patriarchal milieu where this work would have to be in addition to her tasks at home. She may also be expected to hand over her earning to her husband. Yet she could still dream of saving a small amount, if need be unknown to her husband, and look forward to spending a substantial part of the day outside the home. As that hope gains greater prominence within her, she may choose to do something about it. In the pursuit of this hope there would be the suspense of whether she will get a job or not, including the uncertainty around it. This suspense would be in addition to her fears of the still largely unfamiliar environment of the city. A suspense that is dominated by fear could also lead to a response other than escapism. Wimoa could fear the dark environs of her neighbourhood after sunset. This may prompt her to act in ways that reduce this fear, even if this action is no more than taking someone along when stepping out of her home after dark.

The actions in response to suspense are intended actions and could be consistent with those of others that face a similar suspense. The act of breaking a coconut is often carried out in Indian cities as a way of reducing the suspense over the uncertainties about how a project would go. A person's decision to do so would be consistent with those of others who share the same belief. There could be times when the actions to reduce suspense are not consistent with the beliefs of others. The response of one group to the suspense of their hope can come into conflict with the beliefs, or simply the way of life, of others. The religious procession of one group could halt traffic, thereby affecting other groups. The conflict could be more pronounced if the procession involves loud sounds and passes by a place of worship of another group that values silence at that point of time. The incompatibility of the actions of the two groups would be resolved through some form of negotiation. The outcome of these negotiations would be influenced, as in the case of other intended actions, by the power of different groups, the options each of them have, and the sense of fairness they can evoke. In the process the actions that respond to suspense are merged with other intended actions. A group may demand concessions for its religious procession on the basis of the benefits it has the power to offer in other domains. One group could open up options for the intended actions of the other group. The two groups could share some common values which they can evoke in their claims to fairness.

These negotiations need not be explicitly carried out each time there is a potential conflict. The repeated conclusions of a negotiation with the same result would over time institutionalise it. If, for instance, the outcome in the negotiations between the two groups on the religious procession is that the procession will stop its loud music when it passes by the place of worship of the other group, and this conclusion is repeatedly arrived at, it would become the norm for that city. The continued practice of that norm would make it an institution, with or without explicit official sanction. Thus as the city negotiates its way through the actions of individuals and groups − whether intended actions or happenings − it would develop norms that the city would prefer to be governed by. The process through which these norms emerge, and the extent to which they are adhered to, would depend on the nature of the institutions a city develops.

References

Burman, J. & J. Roy, 1996. Hindu-Muslim Syncretism in India. *Economic and Political Weekly* 31(20), pp. 1211–1215.

Dalton, Jon C., David Eberhardt, Jillian Bracken, & Keith Echols, 2006. Inward Journeys: Forms and Patterns of College Student Spirituality. *Journal of Courage and Character.* Accessed October 29, 2018. https://www.tandfonline.com/doi/pdf/10.2202/1940-1639.1219.

Groesz, Lisa, Shannon McCoy, Jenna Carl, Laura Saslow, Judith Stewart, Nancy Adler, Barbara Laraia, & Elissa Epel, 2012. What is Eating you? Stress and the Drive to Eat. *Appetite* 58(2), pp. 717–721.

Haque, Jiaul, 2018. Land, Caste and Power in Circular Migration. Working Paper, Bengaluru: National Institute of Advanced Studies.

Heatherton, Todd E. & Roy F. Baumeister, 1991. Binge Eating as Escape From Self-Awareness. *Psychological Bulletin* 110(1), pp. 86–108.

Henning, Bernd & Peter Vorderer, 2001. Psychological Escapism: Predicting the Amount of Television Viewing by Need for Cognition. *Journal of Communication* 51 (1), pp. 100–120.

Livezy, Lowell W. (ed.), 2000. *Public Religion and Urban Transformation: Faith in the City.* New York: New York University Press.

Majumdar, Boria, 2002. Cricket in Colonial India: The Bombay Pentangular, 1892–1946. *The International Journal of the History of Sport* 19(2–3), pp. 157–188.

Mall, Sangeeta, 2017. In Godmen We Trust [online]. *Australian Humanist* (128), pp. 8–9.

Mildnerová, Kateřina, 2016. The Modern Forms of Witchcraft in Zambia: An Analysis of Local Witchcraft Narratives in Urban Settings of Lusaka. *Religio* 24(1), pp 19–51.

Ortony, Andrew, Gerald L. Clore, & Allan Collins, 1990. *The Cognitive Structure of Emotions.* Cambridge: Cambridge University Press.

Pani, Narendar & Nikky Singh, 2012. *Women at the Threshold of Globalization.* New Delhi: Routledge.

Vorderer, P. 1996. Rezeptionsmotivation: Warum nutzen Rezipienten mediale Unterhaltungsangebote? *Publizistik* 41(3), pp. 310–326.

Yanal, Robert J., 1996. The Paradox of Suspense. *The British Journal of Aesthetics* 36(2), pp. 146–158.

7 The making and unmaking of institutions

The actions an individual pursues in a city range from the essential one of breathing in air – polluted or otherwise – to more complex intended actions of, say, designing a plane. Her intended actions have to be carried out in the midst of happenings she has no control over, which generate their own suspense. The pursuit of an action can at times be a lonely, individualistic process, as when the person wants to write a personal diary. More often the pursuit could be shared with others who seek a similar, if not the same, action. The sharing of a pursuit may be no more than seeking the joy of doing things together, but it could also be prompted by the need to defend an action against that of others seeking a contradictory outcome. Wimoa's friends include vendors who make a living from the small place they occupy on a pavement of their city in the global South. Their occupying this place on the pavement typically leads to pedestrians having to walk on the streets, which in turn reduces the place for moving vehicles. The vendors' action of occupying the pavement brings them into conflict with those seeking more place to drive through the street. When push comes to shove, the street vendors would try to mobilize whatever little support they can find, as would those seeking to have them removed. The process of this negotiation, often leading to conflict, would determine not just the outcome of that specific dispute but also the larger relations in the city between those who see the vendors as one of their own and those who have a similar association with those who drive cars.

This little story of vendors, similar to what vendors face in several mega-cities of the global South (Bhowmik, 2005), provides a sense of the everyday negotiations of a city. An individual who negotiates her way to pursuing specific actions, making allies and opponents on the way, participates and influences the negotiations between these groups and hence social relations of the city. The negotiations reflect her influence, howsoever miniscule it may be, and the collective weight of her group vis a vis that of others in the city. The ability to influence the negotiations would be enhanced if she and her group have options that others do not want them to exercise. A street vendor who is evicted from her place may have the option of moving somewhere else. This would inconvenience those in her existing neighbourhood who are her regular customers, possibly leading them to support the street vendors' cause. A person would thus negotiate her way,

DOI: 10.4324/9781003196792-7

sometimes successfully sometimes without much success, towards carrying out her actions, whether she does so individually or as a part of a group. She would be a part of the multitude of negotiations between individuals and groups that form everyday life in the city. If and when these negotiations throw up the same result repeatedly, they become the norm. There are cities where the negotiations over traffic have led to a norm of all vehicles stopping at a red light, just as in other cities this negotiation has not progressed so far as to ensure that all vehicles follow this norm.

Negotiations and autonomy

Among the reasons why negotiations do not always lead to repeated conclusions, and the resultant establishment of norms, is a perception of unfairness. An individual or group that is convinced, reasonably or otherwise, that they have not been adequately represented in the negotiations could decide they have little reason to follow the norm. This alienation is enhanced by the many inequalities that could, and do, impact the negotiations of a city. The more widely recognized of these inequalities are the economic differences between individuals and groups in a city. Economic power can influence the course of negotiations in ways that consolidate existing inequalities. Excessive economic power in the hands of a few typically leads to the skyrocketing of land prices in the city, thereby contributing to the poor being priced out of the market for basic housing.

The power of economic inequality in the processes of urban negotiations is not confined to differences between the rich and the poor. There can be considerable antagonism between different groups of, what would usually be generalized as, the rich. This antagonism can be quite intense when a city goes through a funda-mental economic transformation, replacing one elite with another. The commu-nications revolution saw many corporate houses, headquartered in what has been termed global cities (Sassen, 2001), tapping technical manpower resources available in cities in what was then considered the developing world. As has been noted earlier, these circuits allowed for companies to be set up in some cities of the global South to consolidate manpower resources and use communication technology to link them directly to the demands of command and control centres often located in global cities. The success of these, sometimes technologist-led, companies con-tributed in no small way to the countries in which they were located being reclassified as emerging economies from their old status as developing economies. The magnitude of this transition raised the profile of the companies involved and the individuals associated with them. In India, Nandan Nilekani, a founder of one of the information technology companies, Infosys, went on to create one of the most extensive, and some have argued intrusive, identification systems – the Unique Identification Authority of India (UIDAI) which generates Aadhaar cards that provide biometric identification for Indians (Mukherjee & Nayar, 2011). The new elite was also given a prominent role in cities, often being asked to lead initiatives well out of their areas of expertise. Nilekani headed the Bangalore

Agenda Task Force a body set up to advice the state government on matters related to that city (Ghosh, 2005). The interests of such new elites in a city need not be, indeed are usually not, in sync with those of the older elites.

Economic negotiations may concern interests in a particular city but the factors influencing that negotiation are not always confined to that city. The ability to bargain of a manufacturer in Chittagong in Bangladesh providing garments for a global brand would depend on conditions on the other side of the globe. The command and control centre of that global brand, located in a global city, would decide whether this manufacturer is the best option available. The negotiations of the manufacturer with the global brand would, in turn, influence the negotiation between the manufacturer and her workers. The outcome of that negotiation would go on to influence the course of the workers negotiation with those from whom they buy the commodities they consume. The series of negotiations deeply influence the future of the manufacturing unit in Chittagong. The bargaining position of the parties at each stage of the negotiations would depend on their resilience and the options they have. In deciding whether to continue with that garment manufacturer, the global brand would consider its other options, whether in Chittagong itself or in any other potential manufacturing centre in the world. The ability of the original manufacturer to bargain would be enhanced if she has other potential buyers. The bargaining position of the workers too would depend on whether they have the option of employment elsewhere. If Wimoa were to be a worker in this factory in Chittagong she would not know very much about the negotiations between the global brand and the manufacturer, though she would potentially be among the worst hit if the factory were to close down. The mixture of rumours and gossip she has to make do with would usually ensure a suspense which includes the hope of better wages alongside the fear of loss of the job.

Social negotiations too can be distorted by inequalities, perceived and real. The social elite of a city, with its access to effective networks, city politics and economic power, could have an overwhelming influence on the social norms a city sets for itself. A powerful and exclusionary social elite could contribute to the consolidation of existing social inequalities, whether they are based on race, other ethnic identities, or religious identities. These social relations are not immune to change. The change can be dramatic, as in the case of the storming of the Bastille in the French Revolution, or much quieter, as when old prejudices fade away, making way for new social norms. Wimoa may come into the city deeply rooted in the prejudices of her village; prejudices that prevent her from associating with certain social groups. As she tries to negotiate her way through the vast numbers in the city some of the prejudices may fall away. She is likely to face fewer restrictions in the city on who she can talk to. Her friends from more traditional villages, who have moved to a relatively less traditional city, may find they no longer have to cover their faces. The decline of old prejudices does not necessarily imply the decline of prejudice *per se*. Very often new prejudices take their place. The flip side of the anonymity that the city provides is that one section of the city would know very little about the other. The suspense about what the other may be doing generates its own prejudice. Fears of crime can lead to suspicions of others who are not known. This suspicion can be easily stereotyped into particular classes or communities, leading to prejudice. Urban negotiations between these groups would have to cross this barrier of prejudice.

The negotiations in the cultural domain can be more contentious. The shared and learned behaviour, to follow Margaret Mead's view of culture, would vary from group to group. The behaviour of each group would itself have been negotiated within that group, influenced by the past of its members as well as their sensitivity to change. The negotiations between groups would, in turn, be influenced by whether the dominant mood in the city is to allow diverse cultures the autonomy to celebrate their differences or whether the mood is to enforce cultural uniformity. Much then depends on the attitude to difference, an attitude that can be tested around a range of differences across a variety of domains. This is evident even in the limited domain of cultural performance. At the heart of these performances are differences between the performance and the audience, with the audience expected to appreciate performances that they themselves cannot create or perform. There is also the perceived difference between performers, with one being more appreciated than another. The competition can get more intense when the appreciation is commercialised and then further amplified by a star system that concentrates its monetary appreciation on the few at the very top. And then there is the potentially volatile difference between audiences. The existence of diverse audiences seeking varied performances can enrich the cultural environment of a city, just as efforts to impose uniformity in the tastes of audiences can severely limit cultural growth. It lays the ground for conflict between groups seeking uniformity for contradictory cultural views.

Whether cultural, or any other, conflict in a city slips into violence also depends on the extent of autonomy available for individuals and groups. In an atmosphere of extreme individual autonomy, bordering on anarchy, the householder may decide to ignore the concerns of the city at large. This could lead to a city where the norms of cleanliness followed within the household are not extended to the city as a whole. Indeed, in their efforts to keep their homes clean they would not hesitate to throw their garbage on the streets leading to the potential for further conflict. Extreme individual autonomy could see citizens pushing each other out of the way to get into a bus. Much as extreme individual autonomy can harm the ability of individuals to live in a group without resorting to personal violence, autonomy for a group could have the opposite effect on violence in the city. Autonomy for a group allows it to practice its own norms, just as other groups practice theirs. The autonomous space for each group reduces the potential for tension between groups, tensions that could lead to violence. Ideally, individuals in a city would give up elements of their personal autonomy to follow social norms, even as groups are given the autonomy to follow their own norms. When attitudes to individual autonomy and to group autonomy are the same, the result may well be increased conflict. When the city reduces both individual and group autonomy, it would lead to disciplined individual behaviour, but the efforts of one group to enforce that discipline on another could emerge as a formula for urban violence. Cities like Mumbai usually find their citizens quite willing to make way for each other in the course of everyday life, but the same metropolis has been known to reach levels of murderous violence at times of Hindu-Muslim confrontations. The picture is, if anything, worse when a city offers high levels of both individual and group autonomy. As individuals pursue their interests without a care for others it can lead to considerable inter-personal violence, with discipline only being enforced by those who control a group. Those who control a group can then direct the inherent everyday violence of its members against other groups, creating a platform for locally organized riots. The acceptance of violence in the everyday life of a city thus depends to a considerable extent on the autonomy that is available to its citizens and groups.

The levels of autonomy that are available to individuals and groups are themselves negotiated. When the negotiations are dominated by the power of one group, with its own sense of what is fair, leaving other groups with few options, there would be a tendency to enforce the norms of the dominant group by eroding the autonomy of others. In negotiations that are marked by more evenly matched groups there is a greater likelihood of the negotiations resulting in greater autonomy for individual groups. The stability of the arrangements in the city would depend on the wide acceptance of the results of the negotiations. The acceptance of a result is, however, no more than the best that each individual or group believes they can get. And the best they can get would keep changing over time. The power and the options of the participants in the negotiation are not static. Economic, social, cultural and other resources could change, especially when the city becomes the destination for new processes of agglomeration and polarization. The negotiations that determine the course of actions in a city are thus in a process of continuous flux, where both the players and the outcomes can keep changing.

The making of institutions

The negotiations that mark the many and varied actions in a city can influence the everyday life of its citizens not just through their outcomes but also in the way they are carried out. Ideally everyday life would be marked by a set of outcomes for which explicit negotiations are not necessary. Taking a taxi in a city and paying the metered amount at the end of the trip would be a case of the outcome being arrived at without an explicit negotiation. Such a clearly laid out basis for negotiating an action is not always available. There are cities where the amount to be paid for a taxi has to be negotiated for every trip. The absence of a clearly laid out practice in lieu of repeated negotiations can be disruptive of the everyday life of a city.

The processes of negotiation also involve defining the rights and duties of the participants in that negotiation. A passenger would have the right not to travel in a taxi demanding what she thinks is too high a fare, just as she would have a duty to pay the agreed fare at the end of a trip she decides to take. The processes of negotiation can themselves become negotiable in rapidly growing cities. A new entrant to the city may not know the language of its taxi drivers. The choice of language for the negotiations could itself be part of the negotiations. In cities going through intense conflict between older residents and new ones, the older residents could insist on negotiations in their language.

The acceptance of the result of a negotiation also decides the nature of everyday life in a city. A city could arrive at a set of traffic norms that it follows. These norms need not necessarily be what is laid out in the rule book; they could just be those that are accepted by the drivers of vehicles in the traffic and by the pedestrians. It is this acceptance of the norms that have been arrived at which allows for the smooth movement of traffic. If one driver breaks the norms the entire traffic could be disrupted. The consequences of that disruption would be further negotiated, ranging from ignoring it to an angry argument and road rage.

A stable city would set out further practices to regulate what would be done if existing norms are not followed. It would have a system of regulation that lays out the norms as a set of rules that are to be followed. The failure to follow these rules would be met with penalties. Breaking traffic rules may result in no more than a fine, but breaking other rules of existence in a city could have more serious consequences. In the process the city develops a series of institutions that govern its continuing negotiations.

The concept of institutions has been defined in a variety of ways and has been the subject of extensive debate. In finding our way through these diverse interpretations there is much to be said for the perception of Rawls that

> An institution may be thought of in two ways: first as an abstract object, that is, as a possible form of conduct expressed by a system of rules; and second, as the realization in thought and conduct of certain persons at a certain time and place of the actions specified by these rules.
>
> (Rawls, 2000, p. 55)

He goes on to add that "An institution exists at a certain time and place when the actions specified by it are regularly carried out in accordance with a public understanding that the system of rules defining the institution is to be followed" (Rawls, 2000, p. 55). In a city that is identified with its actions this definition has the advantage of adequately recognizing the roles played by both rules and actions. Where it does appear somewhat rigid is in the suggestion that an institution would cease to exist once its rules are not followed. In a city where actions are continuously being negotiated, challenging the rules need not be rare, and may even be a regular practice. To take a view that at this point the institution ceases to exist may not be entirely accurate. Cities with chaotic, rule-breaking traffic nevertheless do have their institution of traffic police. This institution may be weak, corrupt and ineffective, but it exists and influences the course of actions in that city. Every time a member of the traffic police takes a bribe it is a sign of the weakness of that institution, but to the extent that that action also influences the nature of traffic in the city, it cannot be said that the institution no longer exists. The relationship between rules and practice is then a dynamic one, where the very nature of an institution can continuously change. Changing practices would ideally lead to the creation of new rules, but it can also lead to no more than a willingness to break the existing rules. Understanding institutions in a city identified by its actions would then require a more dynamic conceptualization of institutions. Drawing from Rawls, though in a way that would fundamentally alter his meaning, we can see an institution as existing *in a particular form* at a certain time and place when the actions specified by it are regularly carried out in accordance with a *public understanding of how the rules defining the institution are to be followed.* Such an institution is continuously changing but is still seen as representing a set of rules and practices.

The working of these institutions depends to a considerable degree on the relationship between practice and the rules. The city would have a relatively stable institutional arrangement if the rules emerge from, and remain consistent with, practice. This would make it easier to give an acceptable meaning to the institutional structure. In a city where drivers concede the right of way to pedestrians there could be wider acceptance of a discourse around the need for pedestrians to gain precedence. In contrast, in a city where the rules have not emerged from local negotiations, even the most reasonable norms need not be acceptable. In cities where motorcyclists ride on pavements, it would be difficult to generate a widely acceptable meaning to the simple consideration that the pavement is for pedestrians.

The distance between practice and rules can be widened when the latter is determined by non-local practices. The role of non-local rules tends to be underestimated when the role of governance of a city is seen as no more than the enforcement of rules. This approach usually results in the hierarchy of rule-making institutions being seen only in terms of levels of government. But it is quite possible for external agencies to determine rules within a city. This is evident if we return to our example of the establishment of labour and environment standards in factories for the manufacture of garments for global brands located in cities of the global South. The initial effort to set these standards was through negotiations between governments, particularly in the World Trade Organization (Rollo & Winters, 2002). But when the WTO ministerial in Seattle in 1999 collapsed there was no further serious effort at generating an inter-governmental consensus on the issue. Instead the focus shifted to global brands enforcing labour and environment standards on their own, in order not to risk the ire of their customers in the West. Since the manufacturers were completely dependent on the global brands to buy their garments, they were willing to strictly follow the rules that were laid out on labour and environment standards. The pressure to follow these norms being specific to the garment export industry, there was minimal spill-over of these norms to the other manufacturing units in the city. Yet it did demonstrate the possibility of rules being enforced by those who did not reside in the city (Pani & Singh, 2012).

The possibility of rules for a city being determined outside its boundaries exists within countries as well. There are countries where national norms are laid out for all cities in that country. It is not necessary that each of the cities in the country would have arrived at the same norms if there were to be a local negotiation. A city with a stronger environment movement may be eager to adopt more stringent pollution standards than a city that has no similar influence. If the national rules follow the stringent prescriptions of the environment conscious city, they would be well out of sync with what the environment insensitive city would have arrived at on its own.

The fact of a city's rule-making processes occurring within its boundaries does not automatically mean that the rules that emerge would be consistent with the negotiations of the groups within the city. The rule-making bodies could be controlled by groups that are not in touch with the everyday negotiations of the city. The bureaucracy of a city could be inclined towards enforcing rules that they perceive to be valid, while a negotiated consensus within that urban space could be inclined towards a different set of rules. In an effort to increase the speed of traffic, a city bureaucracy may decide to widen roads by eating into space for pavements. It is quite likely that if all the residents of the city were to negotiate this decision, this trade-off may not go through. This is, no doubt, a case for greater democratization of rule-making in a city. But it must be kept in mind that democratization alone does not automatically rule out the possibility of rule-making being divorced from the everyday negotiations of the city. In electoral democracies a candidate for a rule-making office can come to power on the basis of a set of promises on specific issues. The office itself may, however, make rules not just for the activities related to her promises, but for a variety of other areas as well. In these other areas, the person elected to the rule-making office could be quite out of touch with the everyday negotiations of the city. The mayor of the city in the global South that Wimoa has moved to could have been elected on a platform of making the city more in line with the high technology features of global cities. Such a mayor need not, indeed is very unlikely to be in touch, with the everyday negotiations that concern the lives of those who still collect their drinking water from a public tap.

The distance between the rule-maker and those who are expected to follow the rules can distort the discourse around rule-making in the city. There are several elements of the rule-making discourse that could be used to justify a distance between what the rule makers seek and what local negotiations would have arrived at. It is not unusual for rule makers, particularly in cities of the global South, to speak of global "best practices". There is an implicit view that the cities of the developed world are advanced because of their practices. The recreation of these practices would consequently help the cities of the global

South reach the standards of the advanced cities. The difference between conditions in the two sets of cities are then used as a carrot to gain wider acceptance of the transfer of the "best practices". The possibility that the differences between the two sets of cities would ensure that the practices in a city of the global North would not necessarily work in a city of the global South, is not given too much consideration.

The continuous interaction between rules and practice can lead to the making of stable institutions as well as their unmaking. In periods when rules and practice converge the institutions would be stable. Conversely, in times when the rules and practice diverge the institutions would face periods of instability. This divergence could occur due to a change of rules, as often happens when there is an effort to use rules to change practice. This typically takes the form of governance reform from above where the law is first changed and is expected to be implemented in a way that would force a change in practice. The divergence could also occur when there is a change in practice, as happens when a large inflow of migrants creates new negotiations within the city and, sometimes, new norms. The dynamic nature of both practice as well as the formulation of rules ensures that institutions themselves are continuously changing, demanding varied responses from individuals and groups who are expected to follow the rules.

Institutions and the Proximity Principle

Wimoa's response to institutions in the city would be guided by the Proximity Principle. Her tendency to limit herself to her immediate surroundings is, if anything, likely to be strengthened by her trying to find a place in a city she has not been previously associated with. She may spend the early weeks or months of her stay in the city only relating to those she has to necessarily deal with, whether at home or at work. Over time she may gain the confidence to associate with larger groups in the city, like the vendors on the pavement near her home. The vendors would in turn be concerned primarily with their immediate surroundings. Their conflict with institutions of the city could be the result of a desire to alter their immediate territorial surroundings. To this end, the vendors would mobilize others who share their immediate surroundings. They could gain further support from still others who support their right to be represented in that space. As a vendor mobilizes support in her immediate surroundings in multiple spaces she would be, directly or indirectly, with those who would stand up for her and other vendors. Once the group is formed they would be in a better position to negotiate with others, with each side being guided by the needs and strengths of their own immediate surroundings. These negotiations would result in practices that would alter the nature of the institutions governing that part of the city.

The process through which individuals and groups negotiate with institutions of the city need not always be directly confrontational. Individuals and groups could adopt practices that bypass the rules, such as elite capture of the processes of governance or simply through the illegality of corruption.

Actions and activism

Elite capture can occur at the stage of making the rules itself. In cities where the interaction between rules and practice can be disruptive, there is often a tendency to formulate new rules by stealth. Even when these rules are formally the result of an open discourse, the debates are usually confined to a small group interested in that particular policy. It is quite unlikely that Wimoa and her friends in the city would have the time, and perhaps even the interest, to participate actively in the policy making discourse. More often than not, they would not even be aware of this discourse. Policy makers often prefer to keep it that way. It is usually the city bureaucracy and those in power who work out the framework for the rules a city should follow. If they are wary of the reaction to these rules they could do it by stealth. This would leave those who have to follow the rules in the dark until the time comes to follow them. This imbalance in information has received considerable attention. Economists have been particularly concerned about the problems of information asymmetry when one party in a transaction has more information than the other. Apart from distorting the transaction it can lead to market failure (Rosser, 2003). The problems of unequal information have been noted in other situations as well, including decisions that have are taken in the realm of international relations.

The possibility of large numbers in a city being forced to follow rules that they are not aware of, and which may not be in their interests, has generated a demand for transparency in urban governance (Bank, 1992). The release of information by institutions of the city is expected to enable those living in it to become aware of the rules that are against their interests. This can be done by making the right to information legally enforceable, as has been done in India (Jenkins & Goetz, 1999). In theory, once everyone has the right to information, every individual would have the same information that the public institution has, thereby preventing information asymmetry. In reality several gaps can still emerge. There is the gap between the information that is released and the information that can be accessed by the individual. This could be a major constraint in cities with low literacy rates. The release of information could take the form of making official documents public, but if the person who needs that information is not literate, the information asymmetry would persist. This problem could be compounded when the information is released online when the affected person is computer illiterate, or has minimal access to computers.

In cases where the target audience is unable to read or understand the information that is released, the information gap has to be bridged by a third party. In the more routine information gaps it may be possible for this bridge to be provided within the family. An illiterate woman who is not able to read the expiry date on a carton of milk she has bought may be able to get the information from her school-going daughter. On other issues that require a more complex understanding, she would have to reach out to others. For Wimoa to address the suspense about the future of the unauthorised colony in which she is living in the city would require information from those who can, not only access the information, but also interpret it. It is here that activist groups find a prominent place in determining the actions of a city.

The role of activist action in a city is closely linked to the nature and extent of the practice of democracy in that urban setting. At a somewhat obvious level, unless a city has a tradition of democratic functioning, activism can be a difficult and even dangerous business. There are cities where journalists are killed, especially when they are associated with activism of some kind. Such extreme actions are usually associated with state or state-sponsored actions. But it is also possible for extremist groups who do not accept a prevailing near-consensus to target and kill activists (Dutt, 2018).

The link between activism and democracy also has a more structural rationale. Most of the models of transparency see the citizens as a homogenous group who can confront the government or the institutions that manage the city. But the city, as we have seen, consists of a large number of quite different individuals and groups continuously negotiating with each other. Much then depends on who represents the citizens as a whole. In a democracy, elected representatives would normally be seen to be representing the citizens of the city. But since these elected representatives are themselves in charge of the institutions that make the rules, they would be the ones about whom information is sought rather than being among those seeking information. Transparency would demand that the elected representatives, or the bureaucrats they now command, provide the information to others. Consequently, any effort to use them to represent the "others" would present a conflict of interest.

The inherent constraints on the elected representatives representing the people at large in the cause of transparency, the representatives of "others" emerge through other mechanisms. They could be activists vocally representing specific interests. Mobility activists could demand information on a proposed elevated road funded by the city; anti-corruption activists could seek information on the financial transactions of the city government; and civic activists could demand information on the enforcement of rules regarding, say, noisy restaurants. The specific type of activists that emerge in a city would be related to the issues that the everyday negotiations of the city generate. A city in which its citizens have to regularly face traffic jams could see the emergence of mobility activists.

For activism to be effective it will have to overcome at least three internal challenges. The knowledge needed for activism need not always pass scientific muster. While there are undoubtedly a number of serious activist researchers who would insist on laying out the most rigorous standards for themselves, that in itself is not a guarantee of acceptance by a majority of the citizens. An audience that is deeply affected by corruption may appreciate an activist who reflects their anger rather more than one who makes a more rational case for a way out of the problem. The information that an activist gains would have to be interpreted, and presented, in ways that are acceptable to the target audience. An audience that is interested in the nuances of corruption may be willing to listen to an extended analysis of the problem, but an angry audience would usually prefer more extreme statements of the problem. The demand for extreme measures could be reflected in very high degrees of cynicism about any information that is received.

The effectiveness of an activist could also depend on her ability to rally large numbers behind her cause. This could at times be done by individuals with a personal following, as when film stars rally support for a cause. More frequently, the mobilization would be around the cause the activist is associated with, in the form of specific actions like a protest march. Repeated mobilizations by a particular individual or group would itself take on the character of an institution, with this practice developing a set of rules. The rules would usually pertain to the activities during the mobilization (a sit-in or a march or any other form of protest), the behaviour during the period of mobilization (usually a commitment not to be violent), and the tone and content of the speeches to be made or the slogans to be raised. Cities normally brace themselves for the larger of these mobilizations with practice also determining the places in the city where the protests usually take place.

A third challenge to effective activism comes from the nature and extent of economic resources it can tap. Much as the economic costs of activism tend to be seen as a distraction from the cause it is an essential part of the exercise. The most spartan forms of activism have their economic costs. There are posters and pamphlets to print, transport to help mobilize supporters and banners to be carried. The process of generating these resources can be integrated into the process of mobilization of support. Street corner collection of donations helps raise public awareness of the cause though the amount collected may not be very large. Larger donations would typically be made by individuals who are already familiar with the cause and believe they need to support it. The persons who make the larger donations could also have influence on the other side of the activist-institution divide. They could alter the response to a particular effort at activism. Movements that mobilize fewer numbers than others can, if some of their supporters are influential, get access to greater information than larger movements that lack the same influence on the powers that be.

The power to influence those in power brings with it the possibility of elite capture of activism. The element that lends itself most easily to elite capture is the economic resources activism needs. An exercise in activism may begin by being knowledge-centric, the pursuit of an idea whose time has come. The spread of that idea may, however, be constrained by the lack of economic resources to get the idea across to a larger audience. While the internet may provide a larger audience it still does not reach those, like Wimoa, who are not computer literate and indeed those who are entirely illiterate. The use of the internet is, to use an expression often used in the discourse around open source software, free in the sense of free speech rather than in the idea of free beer. What is more, economic resources can be used to support specific forms of activism over others. This can be done through advertising in the media in support of the ideas of that particular activism. As the media gets more dependent on advertising it could also tend to align its coverage, to varying degrees, with advertising friendly causes.

An elite capture of activism turns the entire case for transparency on its head. Through the selective support of activism it could ensure the increased availability of one type of information over another. In its extreme form, the capture of transparency activism can be used by one section of the elite against another. The type of information that is released can be used to highlight the advantage of a particular project just as easily as it can be used to discredit another. The elite capture of transparency can be used to support a particular project of one section of the elite or oppose that of a competitor. Beyond the intra-elite battles the use of elite capture of transparency activism also distorts, if not overturns, the original purpose of removing information asymmetry. By keeping public attention on the demand for information that suits the elite, it can keep the focus away from the demand for information of relevance to the less privileged in the city. The release of information about elite-friendly plans to make a city in a less developed economy on par with one in the developed world can capture the public imagination in that city. The development of this imagination, and the hope-based suspense that goes with it, can help keep the focus away from the demand for information on, say, what happens to workers who are evicted from an unauthorised colony.

The impact of elite capture of transparency activism ensures the continuation of information asymmetry for the less privileged in the city. Without direct access to knowledge they would be dependent on what activists can do for them. When these activists are overwhelmed by the activism that is not immune to elite capture, they lose their ability to use knowledge and influence in pursuit of the information they need. This would leave them with mass mobilization alone as their weapon in the negotiation for information. This is a weapon that cannot be used too often. For Wimoa, her family, and those she associates with, such protests would involve loss of work for that day, in addition to the possibility of confronting law enforcement agencies. It is a weapon they are only likely to support when there is a great deal of anger about specific rules. In matters that do not provoke such anger, the rules are more likely to be ignored than resisted.

The city can then easily slip into a situation where policy making continues to be largely by stealth and the response of those who feel aggrieved by the resultant rules is to simply ignore them. As has happened in several cities of the global South, there can be righteous rule-making following norms set in the developed world, matched by an equally righteous popular dismissal of the need to follow the rules.

Corruption as informal rule making

The persistence of rule-making alongside a resilient refusal to follow rules, generates substantial room for non-governance within the city. This room for non-governance can be widespread but it should not be taken to be universal. There are areas where the results of local negotiations in the city and rule-making are entirely consistent. In most cities, even those with considerable non-governance, there are typically clearly laid out sets of rules for funerals which are consistent with what the groups in the city believe it should be. At the same time there are areas of non-governance where the rules are very distant from practice. The response to this gap can also vary quite considerably. In cities where there are a large number of such gaps there could be a tendency among those who enforce the rules to focus on the ones they consider important and pay rather less attention to others. This ranking of the gaps between the rules and results of local negotiations could itself be influenced by trade-offs perceived by law enforcers. Law enforcers in most cities in the global South see themselves facing a conflict between the need for economic growth and the concerns of the environment. Until such time that the decline in the environment affects everyday life, as when the smog becomes a serious health hazard, law enforcers may not feel compelled to enforce environmental rules. This has contributed to several cities developing debilitating levels of pollution before law makers resort to desperate measures. At that point the area of non-governance moves from being, in effect, acceptable to law makers to becoming unacceptable.

The period when non-governance in a particular area is considered acceptable by law enforcers lends itself to a variety of illegal negotiations. The law could demand environmental clearances before a particular project is implemented. The process of obtaining these clearances could follow strict rules that have been laid down. But the process could also have the potential for illegality. In a large project any delay can be expensive, and there is a saving in getting prompt clearance. A corrupt law enforcer could decide that he should get a share of this saving and insist on a bribe for prompt clearance. The size of the bribe would vary depending on the size of the savings from prompt clearance of the project. Not all bribes take the form of huge payments for very large projects. There are the much smaller, but more frequently paid, bribes for everyday services. Wimoa may live in a city where the local government issues cards to get the subsidised food she is entitled to receive. The bribe she pays to officials to get the card, and to renew it, is unlikely to be significant in the context of the major bribes paid in a city, though it would account for a large portion of her access to economic resources.

The payment of bribes for services that are perfectly legitimate usually generates a widespread anger against corruption. This anger can be targeted at particular individuals with a reputation for seeking bribes for legitimate services. The defence of such officials is usually that they do not get to retain the entire amount they collect. A substantial part of it has to be paid to those above them in the governance hierarchy who ensured they were posted in positions with a potential to earn bribes. This can contribute to the anger being directed at the entire government leading to protests and even larger movements that overthrow governments. But, more often than not, the change in government does not do away with corruption altogether. It sooner, rather than later, returns through new officials.

To explain this resilience of corruption it is necessary to go back to the nature of the areas of non-governance. In its pure form, non-governance would require rules that law enforcers are not inclined to implement and ordinary citizens are not inclined to follow. Such a scenario would be consistent with areas of complete inactivity. When Wimoa moves into the city she and her husband may stay in a hutment in an unauthorised colony. Lawmakers for a variety of reasons may not be inclined to evict them and the residents do not quite have the option of moving out. As long as this situation continues there is an acceptance of non-governance. But there is an urban suspense underlying this situation. The residents would hope for permanence, fear eviction, and live in a continuous state of uncertainty about their living arrangements. It would make a big difference to Wimoa and her family if their unauthorised residence could become authorised. They would be quite willing to pay a bribe to benefit from the change in the legal status of their home even if that process required going against existing rules laid out for the city. Both Wimoa's family and the official would benefit from this act of corruption and would have to cooperate to carry it out. Since Wimoa and her family are active and willing parties to this corruption it would not be marked by anger against the official, and may even result in gratitude to him when the colony is legalized. The dynamics of such cooperative corruption are thus quite different from that of non-cooperative corruption. The significant role played by corrupt actions in the megacities of the global South makes it worth our while to spend a little time on the distinction between cooperative and non-cooperative corruption; a distinction I have made in an earlier work which I draw on here (Pani, 2016).

The term corruption means different things to different people. There are those who would analyze it as a problem confined to the public sphere, even as others would locate it in other domains, including seeing it as a social pathology. In a widely recognized work Heidenheimer and Johnston (2017) find their way through this wide range of definitions by classifying them into three types: public office centred definitions, market behaviour based definitions, and public interest focused definitions. The public office centred definitions focus on the norms binding on the incumbents of these offices, with corruption being the result of their breaking these norms. Market behaviour based definitions focus on the economic gain the perpetrator of corruption is expected to make, usually based on the theory of the market. The public interest definitions focus on the harm caused to the public by the act of corruption. These definitions may focus on different aspects of the process of corruption but they are not mutually exclusive. It is quite possible for an act of corruption to involve the misuse of public office to make an economic gain in a way that harms public interest. A comprehensive definition of corruption would, in fact, include all three dimensions of the problem. One example of such a comprehensive definition would be Mark Philp's view that

> we have a case of corruption when: A public official (A), acting for personal gain, violates the norms of public office and harms the interests of the

public (B) to benefit a third party (C) who rewards A for access to goods or services which C would not otherwise obtain.

(Philp, 2006, p. 45)

When this definition is placed in the context of negotiation between rule-making and the preference to ignore rules, there are two limitations that need to be overcome. The definition restricts corruption to the actions of a public official. This can be a limiting perception. It is quite possible for corruption to also occur through individuals with no official position. And there is no dearth of such individuals, especially in the cities of the global South. Unofficial advisors have been known to play a role in urban policy in several such cities. In some cases this can be justified on the basis of the perceived expertise of these advisors, while in others there need be no such justification. In Indian cities, and indeed in the rest of the country, it is hardly unknown for relatives of public officials to act on their behalf. It is also possible to be corrupt in entirely private settings. A security guard in a private gated community could, on receipt of a bribe, let in persons who are not supposed to be allowed in as per the rules laid down by the community. Such considerations could have influenced Transparency International's definition of corruption being the "abuse of entrusted power for private gain" (Rubio & Lifuka, 2021). It would be useful then to refer to not just public officials but anyone entrusted with power, whether in government or in the private domain.

The second limitation is that corruption could occur even when all the steps outlined in the definition are not met. It would be an act of corruption when an official violates the norms of public office for personal gain, even if it does not directly harm public interest. A public official taking cash illegally from the office during the evening and returning it the next morning would be an act of corruption, even if it does not alter any public expenditure. Corruption would occur even when a person entrusted with power violates the norms of public office for personal gain without benefiting any third person. It would be useful then to recognize that this definition has at least four separate components: misuse of office for personal gain; knowingly hurting public interest; benefiting a third party who is not entitled to benefit; and receiving a reward for doing so. Each of these elements in itself would constitute an act of corruption.

Each of these four components have been stated in a way that they could cover quite different realities. Whether a particular action is seen as a misuse of office for personal gain would depend on the results of negotiations in the city. In some cases there could be a consensus that the use of an official car for the family is perfectly legitimate, while in others it could be considered illegitimate. To have someone employed by the public office carrying out household tasks of the official could be acceptable in one city and not in another. The implicit negotiations in some cities could allow for actions that go against public interest as when roads are blocked for traffic in order to allow a public official to pass through, while other cities may be loathe to accept such a practice. Similarly, cities that emphasize family loyalty may be willing to accept a family member benefiting when she was not strictly entitled to, while in other, more individualistic, cities this would not be the case. And in cities where giving gifts during festivals is an acceptable, even expected, social practice, city officials may well believe they are entitled to receive these rewards for their service.

The values that emerge from practices in a city need not be consistent with the rules of that city, particularly when the rule-making process is divorced from the implicit negotiations that result in those practices. In situations where the gap between accepted practice and the rule of law is wide, it is possible that there is moral pressure to, in fact, break the law. This tends to happen quite often in cities where the laws are not related to cultural practice. It may be the culture of a city to build houses close to each other, sometimes even sharing a wall. If the rules of that city insist on pre-determined spaces between the walls of different houses, it could be difficult to enforce that law. There may even be a large number of people who feel there is no harm in using a bribe to ensure they can continue with the building norms of their culture. The existence of a large number of such persons could well be an indication of the wide gap between the rules of the city and results of the negotiations within it.

Such a gap may even be used to justify all acts of corruption in the city. A narrative could emerge claiming *all* acts of corruption are the result of laws that are not consistent with the values of the city. Providing such a moral edge to all illegality could, and usually does, lead to a race to the bottom of the corruption pit. It is to avoid this trap that it is important to distinguish between "cooperative" and "non-cooperative" corruption (Pani, 2016). Non-cooperative corruption would occur when those entrusted with power generate illegal private gain from tasks that are legal in themselves. When a city official asks for a bribe to provide a birth certificate a person is legally entitled to, it would be an example of non-cooperative corruption. In contrast, cooperative corruption would occur when those entrusted with power generate illegal private gain by carrying out illegal acts for others. In such situations there could be pressure on the city officials to carry out such illegal acts from those who benefit from that illegality. These illegal acts could develop norms of their own, making them, in effect, an informal system of rule making.

The reaction of citizens to the two forms of corruption will necessarily be quite different. Citizens may have no option but to pay a bribe in a situation of non-cooperative corruption. Someone who has legitimately qualified for a driving license may be willing to pay a bribe to ensure it is delivered to her. But since she is entitled to receive the license, she could see it as an unfair burden. As such incidents of non-cooperative corruption grow in number they can generate popular anger across the city against corruption. In contrast, in situations of cooperative corruption the illegal benefit is shared between the citizen and the city official. It would hardly suit the citizen to protest against this variety of illegal actions. When cooperative corruption is widespread it would be because it is supported by practices emerging from local negotiations. The distinction between cooperative and non-cooperative corruption would explain the tides in the attitudes of cities to corruption. Movements against corruption can at times bring large numbers of citizens to the streets, but there is often little impact on levels of corruption after the movement. This seemingly schizophrenic attitude of cities to corruption would be easily explained if the protests were based on anger against non-cooperative corruption while the continued acceptance, after the movement is over, is of cooperative corruption. Wimoa would find it deeply unfair that she has to pay a bribe from her limited earnings to an obviously richer city official for access to subsidized food that she is entitled to. She may have fewer compunctions about paying the same official a bribe to prevent her and other members of her unauthorized colony being evicted through the enforcement of the law.

The narrative on corruption usually emphasizes the actions of those in government. They are the ones who make rules and enforce them. The implicit assumption that they are the only ones in a city who do so is, however, quite misplaced. The rules regarding the negotiations of the city are worked out at different levels. There are rules, such as the freedom given to a young girl, that are made entirely within a household. The family may also follow other rules set by the local community. The rules for church on a Sunday may be laid out by the local parish just as those for a temple are enforced by its priests. The rules for a company may have to be within norms set for the country as a whole, but nothing stops those in charge of the company from setting rules for practices that the national law does not concern itself with. The rules for a gated community too would be set by members of that community as long as they are within the parameters set by the city government.

The different authorities in rule making for actions in a city imply that there could be corruption at several non-government levels as well. There could be cooperative corruption when a boy bribes his way to avoid punishment by providing his father information about his sister. Wimoa could be a victim of non-cooperative corruption if the supervisor at her factory demands a bribe to record her presence at work. The possibility of corruption, both cooperative and non-cooperative, adds another dimension to the already dynamic nature of institutions. Non-cooperative corruption is likely to be resisted by the person forced to pay a bribe, leading to a weakening of the authority of the institution, and possibly protests against corruption. Cooperative corruption, in contrast, would generate practices that the institution pretends to ignore but is deeply influenced by.

Public and private orientation of institutions

Beyond the deception inherent in both elite capture as well as cooperative corruption, there could be more explicit efforts to take over the process of rule-making. A person would prefer institutions that she believes would protect her interests; institutions that are sensitive to her position in the negotiations. The Proximity Principle would ensure that rather than relying on a distant government she would prefer institutions that are closer to her immediate surroundings. Decentralization of government is an effort to address these concerns, but much of this effort is entirely in absolute space. Decentralization usually refers to a movement from the national government to the regional, and to the local. The Proximity Principle, however, works in other spaces as well. The unauthorized colony Wimoa is staying in could be alongside the campus of a global information technology company. While they are next to each other in absolute space, they are worlds apart in what they relate to, or relational space. Those in charge of the company seeking a global presence would be keen to present a "developed country" version of their city to their collaborators in the global North. Too often their perception of this version of the city includes the removal of the unauthorised colony next door. At the same time, Wimoa's existence in the city depends, to no small extent, on the continuation of the unauthorized colony. A democratically elected local government may struggle to balance the economic power of the company against the votes of the unauthorized colony. Even as the residents of the unauthorised colony seek to mobilize all the political support they can get to remain where they are, the company could seek to insulate itself from the unauthorised colony by building higher walls, within which its own rules would apply. The emergence of such areas of private rule-making has contributed to the debates on public and private spaces.

This debate may need to be rephrased in the context of our earlier discussions on space and place. The definitions of public space often suggest that what is being referred to is closer to what has been termed in book as place. Mitchell (1995), for instance, takes public space to represent "the material location where the social interactions and political activities of all members of 'the public' occur" (Mitchell, 1995, p. 116). If this emphasis on location is accompanied by reference to a locale, a set of actions, memories and a sense of place, it would be in line with our earlier discussion to refer to it as a public place rather than a public space. But that is not the only issue that needs attention. This definition has two elements in it: it is non-exclusionary since it refers to all members of the public, and it refers to the site of social interactions and political activities. The latter element is, however, not confined to public spaces. Social interactions and political activities can, and do, take place in private spaces as well. It is also restrictive in that it does not allow for economic activities, thus keeping markets out of the domain of public space. The absence of a clear demarcation between public and private places for social interactions and political activities may have contributed to the tendency to focus almost entirely on the non-exclusion criterion for recognizing a public place. This is sometimes made explicit, as in the contention that "Public space can in general terms be described as a place open to all, free of charge" (Eriksson, et al., 2007).

Non-exclusion may not, however, be a sufficient condition for a public place. There is a body of literature, going back at least to Jane Jacobs (1961), that takes the argument a step further by bemoaning the decline in public places. The case against this decline is often taken to be no more than a criticism of the private takeover of public places. But in reality, the debate would also extend to the takeover of one public place by another. In rapidly growing cities, especially those of the global South, there can be alternative uses for public place. An open piece of land could be used by children to play, by adults to jog, or by elders to simply sit in the midst of greenery. As the constraint on the availability of land grows in these cities, the competing uses of a public place usually come at the cost of one another: a jogging track at the cost of a children's playground; a children's playground at the cost of quiet open land for the elderly without the risk of being pushed by playing children. Whichever option wins in these negotiations, those seeking the other options are effectively excluded. The ideal public place would then not just allow all to enter it, but would also be one where its use by one set of persons does not reduce its availability for others.

The city being a collection of places there is reason for considerable attention to be paid to place in the discussion of the public and the private. Yet any tendency to confine the discussion of the public and the private to specific locations and their places, can be limiting. The effects of pollution that mark a particular location can, and often does, extend to other locations in the city. It may then be useful to extend the discussion of public places to that of public spaces. Staying in the organic space, public feeding of the poor can have an important influence on the quality of life of the poor, especially in cities of the global South. A discussion on what is public in the multiple spaces a city experiences would then enrich our understanding of the public and the private in specific places.

This extended understanding would demand a sharp conceptualization of the distinction between public and private spaces. In this task we could borrow from the two conditions economists lay out in their concept of a public good. For something to be considered a public good there should be no exclusion of anyone who can consume it (the non-exclusion condition) and its consumption by one person should not reduce its availability for another (the non-rival in consumption condition). The conditions can be adapted to distinguish between public and private spaces. A public space would be one which anyone can enter, and its use by one person will not reduce its availability for another. There would be no economic, social, age, or any other restrictions on entry and all would be able to use the space without in any way restricting its availability to others. A common example economists use is that of air, no living person can be excluded from its use and its use by one does not hurt its availability to others. In a private space neither of these conditions would be met. There would be restrictions on entry, and the use of the space by one person would reduce its availability to another. A private dinner would restrict entry to that organic space to others and when dining by one set of diners it will hurt its availability to others.

As the discussion moves from the locations that are central to place, to the experiences of various spaces, the focus would also shift to the actions that are involved in the experience of space. Whether it is the unavoidable action of breathing in organic space, thinking in the imagination of lived space, or building of a home in absolute space, actions come to the forefront of the discussion. Much of the discussion on the public and the private should then shift to the distinction between public and private actions. A public action would be one that works against exclusion even as it ensures the availability of space when there are already others using it. Allowing entry into the absolute space of a park for all would meet the first condition of non-exclusion. The second condition would be more difficult to achieve since the consumption of any facility that has finite limits would reduce its availability for others. This challenge could, however, be easier to overcome if we move on to relative space-time. It is possible that for a time of the day the availability of a facility in the park is far greater than the demand for it. During this time the use of the facility by one set of would not affects its availability for others. But as we move closer to the peak hour, this could change, and the use by one person or set of persons could reduce the availability for others. A public action is thus situation-specific and not universal. In contrast to a public action, we can define a private action as one that generates exclusion and ensures that the use of a space by one person or set of persons comes at the cost of the availability of that space for others. This is best seen in the absolute space, where a private park that restricts entry to those who pay for it would hurt the availability of that space for others.

Much of the discussion around the public and the private is in the realm of absolute space. Gated communities are among the more prominent of private absolute spaces. The claim to being private is not simply because of the high walls they build around themselves. Indeed, high walls have been used through history as an instrument of public institutions. Ancient fortress cities protected the entire city, its exclusion was restricted to external adversarial forces. For the residents of the region it was supposed to protect, the area within the fortress would meet the non-exclusion clause of a public place. Those fortress cities that were large enough to accommodate all who sought their protection would also meet the second condition of the public place. It would be large enough to ensure that by allowing one set of its citizens in, it did not reduce the place for the other citizens. In essence, the ancient fortress cities were public places. In contrast, the private orientation of twenty-first century gated communities is not difficult to notice. The entry of those who do not belong to that exclusive gated community is closely monitored and the norms of exclusion are strictly enforced with these communities sometimes not even offering parking space to visitors. The second condition of the use of the resources for the gated community reducing the availability of the resources for others is also usually well established. In its most visible form the area occupied by the gated community limits the availability of land for others. The larger the gated community, the greater would be the impact of the gated community on the availability of land for others. And the criteria of rivals in consumption is not restricted to land. It is not unknown for gated communities to tap common resources, like groundwater, thereby hurting the availability of this critical resource for others in the city.

The case for strict norms of exclusion is usually built around a fear of crime. The objective of keeping out crime is sought to be achieved by keeping out those who are suspected of being potential perpetrators of a variety of criminal activities ranging from petty theft to more heinous crimes. As the criminals are not known to the community there is a tendency to extend the fear of crime to a rejection of the other. Typically the other is sought to be defined in terms class. There is an implicit belief that the poor are not known and hence could be potentially criminal. Members of communities that are much less well-off would only be allowed to enter the gated community for a specific declared purpose. Wimoa may be allowed to enter the gated community if she were to work as a maid in one of the houses there, but would not otherwise be allowed entry.

The case of a gated community for exclusion, of those they believe to be the other, need not only be based on class. It could be based on particular lifestyles as well. The exclusion could also be extended to those who adopt a lifestyle that is seen to be below the norms the gated community would like to maintain. The basis for exclusion need not be material conditions alone. To cite a typically Indian example again, a gated community of strict vegetarians in a city in India would exclude all those who are known to eat non-vegetarian food. As the specific norms of exclusion are expanded, gated communities would end up following a pattern of homophily, or the preference for others who are similar. The gated community could then consist of a cohesive group that consists of those who are similar to each other.

The challenge for gated communities is that while they may exercise near-total control over absolute space, this space does not exist in isolation. A gated community also has its lived space, including that of imagination. As noted earlier, a major motivation for gated communities is the fear of crime; they would like to exclude all who they imagine could potentially be associated with crime. Its effectiveness in actually preventing crime is, however, not beyond challenge (Breetzke, et al., 2014). Indeed, there are crimes that cannot be prevented without affecting the very privacy that gated communities seek to protect. This is most evident in the case of crimes within homes in the gated community, particularly domestic violence. While there may be a consensus within the community about preventing extreme forms of domestic violence, this would not always be easy to enforce in cases of violence that are not physical.

The ability of a gated community to manage its other spaces would also vary. The exclusion in organic space can also take the form of restricting membership to the gated community to those who follow specific diets. There are gated communities in Indian cities where non-vegetarian food is not allowed upon, though it is the diet of a majority of Indians. And there are entire cities in the country where the consumption of beef is strongly discouraged, if not banned. In this case the existence of a prescribed diet rules out the availability of alternative diets. The degree of exclusion could make a significant difference to the everyday life of a city, with a city enforcing multiple restrictions on diet being less free than one that allows for varied cuisines.

The actions of exclusion are equally pronounced in the conceptualized space. Cities can be conceptualized in ways that do not reflect the conditions of vast sections of its population. Global cities can be seen in terms of their ability to operate as command and control centres of the circuits of globalization. This characterization of the city would exclude several sections of its population whose connection with globalization is at best tenuous. The use of one conceptualization of a city to override other possible conceptualizations is quite widespread in cities which provide the resources for globalization, with the use of labels like IT (information technology) cities. The overwhelming dominance of such a characterization would not only exclude some sections of the city, but to the extent that it does not leave space in the discourse on that city for other formulations, it also reduces the availability of conceptualized space for others.

The use of conceptualized space for exclusion can occur within cities as well. Sections of a city's population can lay claim to a conceptualization of a part of the city. This is often done through the means of setting up statues of icons who represent that group. Indian cities are particularly fond of this process, with there being no dearth of statues of political and social icons. But the statues of one icon can be brought down by groups who would challenge the acceptance of that icon. The removal of the statues of civil war generals in the United States is an example of this process in cities with disputed interpretations of their histories.

Actions in the multiple spaces of the urban thus generate two sets of conflicts, one between the public and private spaces, and the other between different groups seeking to create private spaces by excluding each other. The first conflict is, arguably, best seen in terms of the air in a city. As has been noted, the organic space that provides the air we breathe is a public space: it does not exclude anyone from the air, and its consumption by one person does not affect its availability to others. But it is affected by actions in private spaces, as in the absolute space of a factory generating polluting gases. The battle to create private spaces also extends to multiple spaces. There could be efforts to exclude others from specific organic spaces, say, in the determination of the diets to be followed in a community. The efforts could also extend to absolute spaces with efforts to restrict the entry of one group of persons to absolute spaces the other would like to dominate. When these absolute spaces have not just a particular location, but also a locale, actions associated with it, a memory it generates, and a sense of place, the battle is one for control over a particular place, including transforming a public place into a private one.

The actions seeking to control a private place, and distinguishing it from a public place can directly influence the course of negotiations in a city. A rather elementary but powerful instrument to dominate the negotiations is to keep some of the stakeholders out of the negotiation. This could be done by extreme forms of information asymmetry where a section of the stakeholders are not even aware of the negotiations to generate rules. When an increase in awareness makes this impossible, private actions can be used to ensure that some of the stakeholders are excluded from the negotiations. The negotiations around the norms for admission of children into government run schools could be bypassed by admitting the child to a private school with its own rules. The practices that are followed in the private school, and the associated rules that enforce the continuation of those practices, are what make that school an institution in the complete sense of the term. On the other side of this divide are those who are excluded from private schools by their economic condition. Their access to education would depend on the existence of government run schools that are non-exclusionary, and are provided the resources that are adequate to ensure that the entry of one more student does not, in effect, reduce the ability of the school to admit others. They would take whatever limited public actions they can to ensure the formulation of rules that ensure these practices of non-exclusion and the availability of resources are consolidated. This would make the government-run school a public institution in the wider sense that the term has been used here.

The course of such negotiations impact the places that go to make a city. The places would be dotted with public and private institutions. Public institutions would seek to create their ideal of public place, meeting both the conditions of non-exclusion as well as that of its use not reducing its availability to others. Conversely, private institutions would work to create their ideal private place in which both conditions of exclusion and non-availability are met. The mix of public institutions and private institutions would result in a range of places in the city, from those that meet both the conditions for a public place to those that meet both the conditions for a private place. There would be a variety of places between these two extremes. There could be a place where the condition of non-exclusion is met, while that of not affecting the availability for others is not. There could be a private place, the occupation of which does not constrain the availability of places for others. In festivals on large open grounds it is possible that the occupation of a place by one vendor does not immediately affect the ability of other vendors to find place for themselves. The extent of this availability could change over time. There is thus an entire range of possibilities between what the public institution would consider the ideal public place and what the private institution would consider the ideal private place.

The conceptualization of the entire range, rather than just the binary of public and private, allows us to explore the variation within what are usually taken to be public or private places. It allows us to distinguish within the public place between a park that caters to the needs of the elderly without reducing the availability of playing grounds for children and parks for the elderly that do limit the playing place a city provides for its children. It allows us to distinguish between private places where the owners limit access, as in a restaurant, and those where there is no access to the public at all, as in a private home. These variations could also change across different times of the day. Private restaurants would be open for some hours and not at other hours. Public places too need not be open to the public through the day. In the longer-term the availability of a place could change. There may be a time when a resource is so abundant that it is not difficult, in effect, to meet the condition for the public domain that the use of a place by one will not affect the availability of that place for another. In a park that is usually empty a person sitting on a bench would not, in effect, reduce the availability of that place for others, but if over time the park gets more crowded this condition would not be met.

Each city would have its unique mix of places created by the actions of public and private individuals, groups and institutions in pursuit of their ideals. As each set of institutions follows its own practice a larger pattern could emerge. These patterns would be unique in that the places of no two cities would be identical. There are just too many negotiations in the emergence of these places for the pattern of one city to be recreated in all its detail in another. But these patterns could influence each other. A private institution could set out to create a place that is based on an idea of private place that has succeeded in another city, just as a public institution could set out to create a place based on an idea of public place that has been celebrated by another city.

Public actions addressing the fears of the non-elite are sometimes seen in the megacities of the global South, in the village within the city. As mentioned earlier, the megacities of South Asia, situated as they are in the midst of a large number of rural settlements, have taken over these villages as they have spread their boundaries. The residents of those villages seek to offset the loss of their traditional economic activities by gaining access to urban land prices. Those with adequate amounts of land have the option of entering into deals with builders to develop the land into apartment houses. These apartments can be sold or rented. The process of tapping real estate profits would involve integrating the village with the city. The land of the village, if it is to get the full benefit of urban land prices, would have to be indistinguishable from the land that is separately acquired and built upon by urban developers. This requires the village to attract as much of the city to it as it can. Far from excluding any section of the city, the effort would be to get nearby urban residents to make the village a part of their everyday life. To begin with, this may be no more than the use of the village market for small purchases, but over time this would go on to considering the apartments in the village as worthwhile places to stay, especially if they are a little cheaper than the rest of the real estate in the city. The action thus reduces the extent of exclusion by seeking to lower the barriers between the village and the city. If the village also invests heavily in the development of apartments to a point where this is in excess of existing demand, the purchase of an apartment would not immediately reduce the availability of the apartment for others. As the village in the city comes up with its own set of practices, if not rules, it becomes another, if more public, institution.

Like everything else about a city the precise set of its institutions can be quite unique. The mix of public and private institutions, the levels of cooperative and non-cooperative corruption, and the extent of the gap between rule-making and accepted practice would all vary from city to city. Underlying this diversity and change, though, are similar processes of negotiations to create accepted practices. The processes of rule-making too can be borrowed by one city from another, even at the risk of it increasing the gap with local practice. The results of the multiple processes of institution making would also be unique for each city, but the broad contours of these processes could have much in common.

References

Bank, W., 1992. *Governance and Development.* Washington DC: The World Bank.

Bhowmik, S. K., 2005. Street Vendors in Asia: A Review. *Economic and Political Weekly*, 40(22/23), pp. 2256–2264.

Breetzke, G., Landman, K. & Cohn, E., 2014. Is it safer behind the gates? Crime and gated communities in South Africa. *Journal of Housing and the Built Environment*, March, 29(1), p. 123–139.

Dutt, B., 2018. Eleven Journalists Killed, 46 Attacked, 27 Cases of Police Action: Report on Press Freedom 2017. [Online] Available at: https://thewire.in/media/eleven-journalists-killed-46-attacked-27-cases-police-action-report-press-freedom-2017 [Accessed 7 January2019].

Eriksson, E., Hansen, T. R., & Lykke-Olesen, A., 2007. Reclaiming Public Space: Designing for Public Interaction with Private Devices. Baton Rouge, ACM Digital Library, pp. 31–38.

Ghosh, A., 2005. Public-Private or a Private Public? Promised Partnership of the Bangalore Agenda Task Force. *Economic and Political Weekly*, 19–29 November, 40(47), pp. 4914–4922.

Heidenheimer, A. J. & Johnston, M., 2017. Introduction to Part I. In: *Political Corruption: Concepts & Contexts*. Abingdon: Routledge, pp. 3–14.

Jacobs, J., 1961. *The Death and Life of Great American Cities.* New York: Random House.

Jenkins, R. & Goetz, A. M., 1999. Accounts and Accountability: Theoretical Implications of the Right-to-Information Movement in India. *Third World Quarterly*, 20(3), pp. 603–622.

Mitchell, D., 1995. The End of Public Space? People's Park, Definitions of the Public, and Democracy. *Annals of the Association of American Geographers*, March, 85(1), pp. 108–133.

Mukherjee, A. & Nayar, L., 2011. *Aadhar, A Few Basic Issues.* [Online] Available at: https://dataprivacylab.org/TIP/2011sept/India4.pdf [Accessed 8 April2021].

Pani, N., 2016. Historical Insights into Modern Corruption: Descriptive Moralities and Cooperative Corruption in an Indian City. *Griffith Law Review*, 25(2), pp. 245–261.

Pani, N. & Singh, N., 2012. *Women at the Threshold of Globalization.* New Delhi: Routledge.

Philp, M., 2006. Corruption Definition and Measurement. In: *Measuring Corruption*. Aldershot: Ashgate, pp. 45–56.

Rawls, J., 2000. *A Theory of Justice.* Delhi: Universal Law Publishing Co.

Rollo, J. & Winters, L. A., 2002. Subsidiarity and Governance Challenges for the WTO: Environmental and Labour Standards. *The World Economy*, April, 23(4), pp. 571–576.

Rosser, J. B., 2003. A Nobel Prize for Asymmetric Information: The economic contributions of George Akerlof, Michael Spence and Joseph Stiglitz. *Review of Political Economy*, 1 January, 15(1), pp. 3–21.

Rubio, D. F. & Lifuka, R. L., 2021. Eyes on 2030: Holding Power to Account for the Common Good. [Online] Available at: https://www.transparency.org/en/blog/eyes-on-2030-holding-power-to-account-for-the-common-good [Accessed 8 April2021].

Sassen, S., 2001. *The Global City: New York, London, Tokyo.* Princeton: Princeton University Press.

8 A city is what a city does

In developing this picture of the city as action, this volume has explored the actions that go into the collection of places that make a city. We have travelled through the urban process to examine the actions that go into agglomeration and polarization. The exploration of the consequences of these processes has included the actions of those who are not directly a part of the processes of agglomeration and polarization but are forced, by virtue of the place they are in, to respond to it. This is most striking in the thousands of villages in the global South that have been enveloped by the processes of urbanization. The collection of contiguous places that go to make a city can also include the occasional place that has insulated itself from the urban process. As the actions of individuals in these processes progress from an urge to act to these urges being fulfilled, they involve negotiations of the person with herself, with supporters and opponents of the action, with groups that would strengthen her ability to carry out the action and those that would oppose it, with the state and other institutions, and with many others. These negotiations can be seen as the actions that lead to particular arrangements, or the activities of the politics of the city. The arrangements that result from this widely dispersed negotiations leave their imprint on the places of the city.

The actions of place

This imprint can be seen in each of the dimensions of the places of the city: the location, the locale, the sense of place, the memories, and the actions. The negotiations around location are often seen in terms of the absolute space, particularly clearly demarcated areas of land. It is these clearly demarcated pieces of land that develop a market value and contribute to the various influences of land prices on cities. The resultant real estate interests are a major influence on the course of many cities in the global South. In several of these cities the higher prices of land act as an effective form of exclusion of the poorest from secure permanent housing. The role of location is also evident in the relational space of a city. The location is often defined in relation to other locations. The location of the command and control centres of specific global circuits are necessarily quite different from that of the resource centres they

DOI: 10.4324/9781003196792-8

hope to tap. Without the different conditions available at each location, there would be no reason to build the global circuit. Within a city, the location of specific places is related to the roles they play in the city. There are places that position themselves along the economic hierarchy of the city, from expensive locations to those that house the poor. Within the cities of the global South, there are the places of the villages that have been enveloped by the urban process by virtue of their location.

The locale of the places that go to make a city works primarily in the lived space. The economic polarization that is done in the absolute space is reinforced by the settings of the lived space. The lived space of the more expensive part of the city must necessarily radiate a different ethos. This effort to make economic polarization evident could be carried out through physical restrictions, as in gated communities, but can also be done, some would say more effectively, by altering the settings of that locality. These settings need not be determined by class alone. As we move down the economic hierarchy it is not unusual for localities to reflect ethnic and other cultural characteristics. In deeply polarized cities, the culturally determined localities could operative with effective boundaries; boundaries that other ethnic groups would hesitate to cross. These unofficial boundaries could often be enforced through fear of the other, reinforcing images of safe and unsafe areas in a city.

The sense of a specific place in a city is typically developed in the organic space. The sight of a striking piece of architecture often tells us far more about a place than words can capture. Most cities would hear the silence of a graveyard, though there are cremation grounds in Indian cities that are sometimes marked by the loud beating of drums associated with some funerals. In either case, the sense of the place is reflected in the sounds associated with it. The sense of touch also provides a greater sense of a place than is often acknowledged. In the crowded, congested cities of the global South, the touch of one person against another is almost unavoidable. Those with a revulsion to such touch could make efforts to avoid physical contact, but they too are responding to their sense of touch. The growing influence of machines in public places, such as in buying tickets for a train, reduces the possibility of the touch of another human being. In the process it develops the touch of fingers running rapidly over keys, a sense that is not as widely available in the rural places of the global South. A place can also relate to the sense of taste through the foods it is associated with. This sense would be accentuated in places that are places of agglomeration of specific ethnic communities, and their foods. The aromas of these foods can sometimes overwhelm a place, particularly in the crowded places of street food in the cities of the global South.

The location, the locale, and the sense of the places of a city all contribute to the memories they generate. The memories of a city are, at the same time, deeply influenced by what a person wants to remember. These memories of a particular place can be reconstructed in the conceptualized space of each person so as to associate a city with specific actions. A tourist's memory of Berlin may be no more than the Brandenburg gate, and the action of having visited it. For someone in the local community who has lived through the period of the Berlin Wall, the memories are bound to be associated with actions of both distress over the years when the Wall existed, and of the joy of it coming down. The close association of memories and actions personalizes the experience of places, ensuring a city is what a city does to the persons who interact with it. A city is not merely a location marked in absolute space, but a more complete experience of persons who interact with one or more of its collection of places. It is quite possible for two persons to argue passionately about two very different views of a city, simply because they have interacted with very different places in it. A global business leader could arrive in the modern airport of a city of the global South, use its expressways to reach an architecturally modern campus, stay at the hotel of an international luxury chain, and fly back the next day. Her experience of the city would be very different from that of Wimoa. Wimoa would have to deal with places the global business leader would never visit, or perhaps even see. The differences would exist not just across persons, but also in the experience of the same person over time. It is not very likely that the locale of the place that a person visits at one point of time would be the same when she visits it the next time. And the person's appreciation of that locale may also change in that time. The places of a city are thus continuously changing, as are the people who interact with it. To adapt the ideas of Heraclitus from two-and-a-half millennia ago, no one steps into the same city twice, the city has changed as has the person's experience of its collection of places.

The person and the place

To seek an unchanging model of a city amidst the continuous flow of the processes in its places, interacting with a variety of persons over time, would be futile. The flux of continuous change cannot be forced into a rigid model that would be true in all situations at all times. On the contrary, even among the primarily resource providing cities of the global South, indeed even within individual countries, there are a variety of patterns that emerge. In India, the dynamics of a city with a prominent place for religion, like Varanasi, is very different from that of Bengaluru, a city driven in recent decades by the communication revolution. The method must thus be one that not only resists the temptation to present all cities as reflections of the same model, but must also provide a larger enough canvas for completely new patterns to emerge. The search for comprehensiveness would then shift from large all-explaining theories, to more complete explanations of specific situations that emerge in the course a city takes. This method would call for a consistent exploration of places and processes in individual cities leading to specific insights. Such explorations could throw up the occasional model, but the greater value of this method would be in its contribution to understanding the diversity, and occasional commonalities, in the interactions between persons and places in cities.

An example of such an exploration could be drawn around Wimoa, through her interaction with the places of the city she has just moved into. At each step she could find herself overwhelmed by one or the other dimension of the places of the city. Her immediate concern would be in the absolute space that she, and her husband, would have to find to live in the city. The location would have to be at a place in the city that she and her husband can afford. This would necessarily be at the lower end of the economic hierarchy of land in the city. The difficulty in finding such a place could be overwhelming, even leading to her having to share a home with someone else in her social network. She would have no illusions about her being at the anywhere other than the very bottom of the city hierarchy, but the locale could still take some getting used to. She would be mentally tuned to the fact that this place in the city would be smaller than what she had in the village, and she would certainly not have around her the vast open spaces of the village. The sense of congestion would extend to the other spaces of her everyday life as well. She would be surprised by just how quickly a street she is walking in could turn into a point of commotion, and just as quickly return to its normal congested self. Her experience of each of the dimensions of the places of the city would develop her conceptualization of the city and the memories she would take with her.

The overwhelming influence of each of the dimensions of place on Wimoa would lead her to intuitively focus on her immediate surroundings, in line with the Proximity Principle. She would seek to carve out her immediate surroundings in each of the spaces of her experience. She may desire an absolute space within her small home that she could call her own. She would be aware that this space would only be available to her for a particular time of the day. Her relational space would determine who she shares the spaces of her home with. She may seek an organic space that allows her to stay with the food she has grown up with. She would like to create a lived space around her imaginations of what a home should be. And she may seek to conceptualize the city in a way that she is comfortable with.

Her efforts to carve out her immediate surroundings in each of the spaces she experiences would bring her up against the overwhelming influence of the places of the city. The location of her home in the poorer localities of the city with small residential areas would severely limit the scope for her personal absolute space. The locale would include the congestion she would never have experienced for long in her village. She would have little, or no, scope to carve out immediate surroundings with access to open spaces anywhere comparable to what her village provided. The place would also ensure that each of her senses would have to deal with completely new experiences: the touch of crowds; the noise of, in Simon and Garfunkel's words, "people talking without speaking"; the taste of food she had never known; the smell of different kinds of pollution; buildings that, at first sight to her still rural eyes, reached up all the way to the heavens. She may choose to reconstruct the overwhelming places of the city into memories that others in her village would admire. Her response to each of these interactions with the places of the city would not leave the places entirely untouched. If nothing else, the impressions of urban places she conveys back to her village, could encourage greater migration to the city, with the resultant increase in congestion. Her personal efforts to influence her own immediate surroundings would generate actions that would, when seen in isolation, be too small to directly alter the places she interacts with, but she adds another drop to the ocean of change that often floods places.

Others in the city would have greater access to economic and social capital to carve out more substantial immediate surroundings than Wimoa. There would be those who are powerful enough to change the entire locale of a place through gentrification. But even these powerful individuals would have to deal with the responses of the collection of places that make a city. The gentrification of an area could still leave it affected by the contrasting locales of neighbouring places. This could contribute to an effort to shut out neighbouring places by developing gated communities in the gentrified place. This would contribute to the stark contrast between neighbouring places in a city, a picture that emerges in many cities of the global South. The impact of the actions of individuals at the upper end of the economic hierarchy of the city may be more visible than those of Wimoa, but they are part of the same process of individual efforts to create their own immediate surroundings within the larger processes reflected in the places of a city. The individual actions would influence the nature of the place. The magnitude of the effect of these individual actions may vary very substantially between Wimoa and the person who can gentrify an area, but in both cases the effort to influence their immediate surroundings leaves a mark on the places they interact with.

The overall impact on the place would be mediated by a variety of other elements. Wimoa's miniscule individual ability to influence a place would be magnified manifold if it is a part of a collective action with a larger number of similarly placed persons. The powerful initiator of gentrification may find she has to deal with the actions of a number of her competitors. The pursuit of each person's immediate surroundings would require negotiations with others, especially those who believe they have a stake in the same place. The places of a city would have to continuously deal with a vast variety of negotiations. The participant in these negotiation would vary from situation to situation, those involved in influencing the real estate prices in a place may not necessarily be directly involved in the negotiations within a home on what the well-dressed young lady should be wearing. These negotiations could be formal, as in the case of the process of making land policy in the city, or they may be entirely informal, as in the response of a parent to the tantrum of a young child seeking to visit a particular place like a toy store. The relationship between those involved could be transitory, as is usually the case in the negotiations around an incident of road rage, or it could longer lasting, as sometimes happens in feuds between business families in some cities of the global South. The course of these negotiations could leave their mark on the places they interact with. A negotiation that goes terribly wrong at a particular place, leaving a trail of violence, would affect the sense of that place, the way it is conceptualized, and the way it is remembered. The possibility of violence is a reminder that the actions during the course of a negotiation, and not just the outcome, affect the nature of a place. What is done in the course of negotiations thus interact with a place, affecting its locale, the sense of the place, and memories of it. A place is then also what is done by those who interact with it. As a collection of contiguous places, the city is then what it does.

Actions and diversity

Understanding a city in terms of what it does, provides its own responses to the challenges facing urban theory. As was noted at the outset of this volume, one of the major challenges to urban theory is that of dealing with diversity. Typologies of cities tend to focus on what is considered to be a primary feature of a city, thereby underplaying its other features. This tendency is particularly evident when a few criteria are used to create entire hierarchies of cities, like World cities. The limits of these hierarchies have contributed to approaches that celebrate the diversity of cities. The concept of Ordinary cities brings to the fore the fact that no two cities are entirely identical. But this does not mean that the experience of one city is completely irrelevant to the course taken by another. The challenge is then to celebrate the unique character of each city even as we recognize what one city can learn from another. Viewing cities through the method of actions allows us to do so. It is possible, even likely, that the actions and their consequences in one city could work similarly in similar situations in another city. The mobilization of workers around wage demands could work similarly in different cities, even if the processes are not entirely identical. There may then be lessons generated in one such experience from a particular city that may be relevant in another city. This does not mean that the experience of the mobilization of workers would be identical across all cities, and much less that the relationship between the mobilization of workers and other actions would be exactly the same across cities, or even in the same city at different points of time. On the contrary, the process of mobilization of workers and the context in which it occurs would be unique to each city. A focus on actions ensures that we can learn what each set of actions has to teach us, even as the combinations of actions in all the places of the city brings out the uniqueness of each city.

The ability of the method of action to explore difference helps it distinguish between elements affecting a city, and the city itself. This is best seen in the emphasis it lays on the distinction between the urban and the city. The urban consists of the processes of agglomeration and polarization, and their consequences. The city is a collection of contiguous places. The urban process expresses itself in specific places. These place could be located in multiple cities, and, where agglomeration is in response to circuits of globalization, these cities would be spread across the world. A process of agglomeration may be so overwhelming in the public mind as to identify the city entirely with that particular urban process. It is not unusual for cities in the global South, with places that are sites for agglomeration prompted by global information technology related circuits, to be referred to as information technology cities, or variations of that name. Yet these cities have many places that are not, particularly not directly, related to information technology circuits. To reduce the entire collection of places in a city to a role played by a particular place, or even set of places, is to diminish the city. The distinction between the urban and the city that the method of actions emphasizes thus militates against identifying cities in terms of their role in particular circuits, such as global cities or (to cite something I have been guilty of) resource cities (Pani, 2009).

The distinction between the urban process and the places of the city is also reflected in the seemingly contradictory experiences of the global and the local. A focus on actions makes it clear that the global and the local may, in many ways, represent two different worlds, but in terms of what a person does, the two worlds not only coexist, but could also feed off each other. The people of a city are often a part of global processes even as they remain rooted in the local of everyday life. The processes of agglomeration that bring them to a city can be the result of global circuits. These circuits, and the agglomerations they generate, have the potential to link the most remote of villages to global brands. Workers can be drawn from remote parts of a country to work in the construction of campuses in cities of the global South, to house the technical manpower serving global information technology brands. Yet the workers constructing those campuses are likely to know very little about the global brands that the technical manpower serves. Their concerns in the city are likely to be much more local, whether it is the congested sheds they live in or the food they have to make do with. The participation of the individual worker in the city in both the global and the local domains would throw up a dilemma of whether we should treat the workers and, by extension the cities they are a part of, as a part of global or local processes. It is tempting to address this dilemma by beginning from one extreme or the other, say from the global of Planetary Urbanization to the local of Ordinary cities, and then find a way towards the middle. But once we move away from the clarity of the abstract extremes, how exactly do we determine when we are in the middle, and whether the middle is, in fact, where we should be. The middle may give the impression of academic balance, but does it tell us all that we want to know about the city? The method of actions finds a way out of this dilemma. By going beyond the individuals to their actions, it deals with both the global actions of an individual as well as her local actions. By treating the urban as the actions related to agglomeration and polarization, and their consequences, it recognises this as potentially a part of global processes. At the same time the actions related to the places of a city are local. The method of actions allows us to recognize the global in the urban even as we celebrate, and remember, all that is local in a city.

The comfort of the method of action with diversity need not be universally appreciated, as there are urban processes that will be better served by uniformity. The processes that generate agglomeration have reason to prefer uniformity. The large technologies that drive the economic engines of agglomeration require workers to carry out the same task in an identical manner across the world. The process of ensuring this uniformity can itself be uniform, as when the same management lessons are advocated in very different cities across the world. The underlying belief that uniformity is an essential ingredient of efficiency can spill over to negotiations outside the workplace. The uniformity can extend to the way people like to live, as in rows of identical houses. The search for uniformity outside the workplace is not free of contestation. The particular pattern around which there is to be uniformity is open to debate, with the ideas of one section of the city seeking to influence, if not dominate, the ideas of another. The negotiations between these groups could see different places in the city following their uniform patterns. There may even be places in a city, especially if they are not a part of economic processes that demand uniformity, which would prefer diversity to the point of being near anarchic. The results of these negotiations between types of uniformity, and between uniformity and diversity, is often reflected in the architectural styles of a city. Cities where negotiations are dominated by a particular group with its own uniform architectural style, can be predominantly characterized by variations of the same style. Cities where the negotiations are characterized by more evenly matched groups, can see each group gaining autonomy over its own part of the city. The city is then characterised by sections having very different and distinct architectural styles. Cities in the global South are often characterized by these negotiations being incomplete. Even as one arrangement is reached between groups in the city, a fresh process of agglomeration can reopen the negotiations. The incomplete negotiations at any given point of time can result in the city following random architectural styles as well as aesthetics, as in the colours of their buildings.

The incomplete negotiations in the cities of the global South can also affect the performance of its institutions and hence the governance of the city. When institutions are seen as a combination of practices and rules, a change in practices would generate a pressure for a change in the rules. A new agglomeration would generate fresh negotiations over the distribution of the resources available in a city. A global information technology circuit would, for instance, generate points of agglomeration in already existing cities. These points of agglomeration would demand a share of the resources in the city whether it is in terms of land for its campus, housing for its manpower, or financial support to improve working conditions. The global nature of this circuit could necessitate working hours that extend late into the night. If the city has rules regarding the working hours of women, it would be in the interests of the circuits that these rules are changed. Until such time that the rules are changed, it may well find ways to subvert these rules. This would, in turn, encourage others who find some other rules inconvenient to subvert them as well. The subversion of rules would offer opportunities for those working in institutions of governance to demand a price for their looking the other way. This process of subversion of rules, with the support of those responsible for them, generates cooperative corruption. In this cooperative process, corruption is considered illegal, but not necessarily immoral. Efforts to remove this corruption through legal processes thus tend to lack the necessary moral drive. An approach based on actions would necessarily recognize the moral dimension of the crisis, in addition to the legal one. It would emphasize the importance of bringing the rules of institutions closer to the moral norms that are negotiated by those involved. It would militate against working out a set of ideal rules in the abstract and then imposing them on a city.

Territories of actions

The possibility of the rules of state-dominated institutions being bypassed by those driving new processes of agglomeration into the city calls for a closer look at the role of actions in determining the territories of the city. An action carried out in a city occurs in a specific context, as a part of a particular process. There are actions that are a part of urban processes of agglomeration, even as some other actions are associated more closely with other processes of the city. And it is quite possible for the two sets of actions to be carried out by the same person. A person working in an office in a process that drives agglomeration would also carry out actions that are unrelated to that process. If her office is located in the global South and provides backend services to a market in the global North, she could spend her working hours carrying out actions of the globalized processes of agglomeration. When she stops to buy vegetables on her way home, her actions could be a part of predominantly, if not entirely, local processes. Her actions break away from all three assumptions of the territorial trap, that the state has total sovereignty over the territorial space, there is a clear distinction between the domestic and foreign aspects of her life, and that the economy and society are within state boundaries. Her actions at the workplace may be controlled more by the norms followed by the requirement of a foreign market, which are more in line with those laid down by a Sovereign abroad rather than her own government. Her continuous, sometimes personalized, interaction with customers in a distant country blurs the distinction between the domestic and foreign aspects of her life. And the economy and the society she is catering to are certainly not within state boundaries. It would be difficult to interpret her everyday life in the city as falling within state controlled territories, even if that territory is defined by the geographical boundaries of the city.

To take a view, on that basis, that the person's everyday life is without territories would, however, not be accurate. There are clearly demarcated boundaries between the place of her work and the place of her home. She cannot quite bring her actions in her kitchen into the workplace. And among workplaces there can be a further demarcation of territories. The actions carried out within a process of agglomeration have to follow the norms laid out for the territory of that specific process, such as labor and environment standards, even if that territory cuts across national boundaries. In fact, if we drop the centrality of the state – or state-like authority – her actions would meet assumptions not fundamentally different from those of the territorial trap. While the state may not have sovereign control over her actions within the workplace, there is some other authority, like her employer, who does. The demarcation between domestic and foreign may be blurred in some of her actions, but there are other actions where it is not. The worker in a backend office in the global South may prefer to speak in her mother tongue, say the south Indian language of Kannada, but when she acts as a part of the larger globalized process she would be required to speak the language used in that process, such as English. Conversely, when she steps into the local vegetable market, she may find that her knowledge of English is not very useful. Her domestic and foreign actions would thus have their own territories. And the economic and social processes of her foreign actions would be confined within the boundaries of her workplace, even as the economic and social processes of her home will be within domestic boundaries. The conceptual distinction between the action and the individual allows us to recognize that even as the authority of the Sovereign may diminish, especially in a globalized world, it is replaced by other authorities that demarcate territories of their own. Those in charge of these territories do not have complete control over individuals, but they can, and do, command their actions. Each territory would have its own means of ensuring its norms are enforced, whether it is the threat of losing a job, or simply one of not being able to bargain effectively in the vegetable market. The city thus has multiple territories of actions demarcated for specific processes or sets of processes. Each territory has its own norms laid out for the actions that are allowed within it.

The multiplicity of territories can leave its mark on the absolute space of cities. When the norms in each territory are very different, each territory would like to present its own norms as that of the city as a whole. The conceptualization of the city would have to be negotiated by groups representing these diverse norms. When a particular group, and the territory it represents, dominates these negotiations, the city as a whole would be presented in a way that reflects these norms. In the absolute space, the buildings and the physical infrastructure would be constructed in ways that are consistent with these norms. A territory that celebrates size, would have taller and larger buildings, connected with expressways. Creating a single conceptualized space for the city as a whole is not always possible. Even if a group is dominant it would need others for its own success. A large financial metropolis would need less skilled workers to construct and maintain it. If the living conditions of these workers do not match the conceptualization of the city in terms of size and glamour, the absolute space of the poor can be hidden in ways that they are not presented to those the city is trying to impress. In a megacity of the global South, long elevated roads can be constructed to pass over its more congested and poorer areas. When these roads effectively link the airport and the elite sections of the city, the casual visitor can be left with the impression of an economically successful metropolis.

The study of the urban

The study of a city is not independent of these impressions. The scholar is a part of an academic territory with its own norms. Her academic actions would, in all probability, be consistent with the set of academic norms that are expected to be followed when understanding cities. An analysis based on actions would recognize the possibility of inconsistency in the actions of individual researchers. In a simple, and somewhat obvious sense, this would be true when we lay out a canvas that includes both her academic and non-academic life. Her actions need not be consistent across the various places she encounters during the course of her everyday life. She may choose to be a deep introvert at her workplace, but gregarious at home. Her actions need not even be consistent across the various spaces of her everyday life. She could take very different approaches to her absolute space, organic space, relational space, conceptualized space, and lived space. She could be very rigid in one space, and very generous in another. She could be very rigid about her diet in the organic space, even as she is very flexible in her views of how the city should be conceptualized. The variety of the individual's actions can also be enhanced by the fact that they may not be based on rationality alone. Her actions can also be influenced by intuition or just popular belief, especially when the continuous change in the city introduces an element of uncertainty.

The variety in an urban researcher's actions does not mean they are random. In the midst of a multiplicity of possible actions, she would tend to fall back on the Proximity Principle. The intuition in this principle would lead her to prefer the spaces of her immediate surroundings within the anonymity of the larger processes in the city. The challenge would be when her different immediate surroundings demand contrary responses. The immediate surroundings of her studying her own community may generate a belief in her that there is an ethical need to highlight individual cases of gross injustice, so as to provoke remedial measures. Her immediate surroundings in academia could, in contrast, believe the appropriate ethical response is to anonymize the affected individual. The demands of the immediate surroundings of academia and that of the urban researcher could vary in other ways as well.

In navigating this difficult terrain in academia and beyond, the researcher would find herself repeatedly evaluating her qualitative judgements. In doing so, she would benefit, to whatever limited extent, from a recognition that statements can involve varying degrees of qualitative judgement. There could be statements where their universal value beyond much dispute, even as there are others that rely heavily on her qualitative judgement. The researcher would benefit from going back to a distinction, made in the introductory chapter of this volume, between models, explanations, and methods. Models, which are abstractions of particular relationships between different actions and their consequences, are expected to work across time and place. The more successful of these models would be widely accepted as true, leaving very little room for subjective judgement. Even when there are a large number of models available it may be difficult, though, to find one that will explain every aspect of a situation in a city. They would usually be better suited to explain very specific, and limited, parts of that situation. The researcher could choose to piece together different models, but even that picture is unlikely to be complete. The researcher would then have to complete the explanation of a situation by developing her own insights. She could arrive at these insights from various sources, involving both quantitative and qualitative data. This book has argued that the quality and persuasiveness of this explanation would be enhanced by the method of action. This method would explore the city in a detail that an approach based on individuals or groups would not. It would bring to the forefront the negotiations that extend to a vast mass of actions in a city and in urban processes. These negotiations can be within institutions or on the streets; they could be simple short-term acts of forming groups before crossing streets in unruly cities or longer-term negotiations about who governs a city. The method of action allows for a perception of the city as a collection of places with multiple processes. As continuous negotiations keep altering the nature of these processes, the method of action is sensitive to change. The change could begin with the individual, extend to groups, and then influence negotiations that influence multiple places in the city. A persuasive understanding of the city would not lose sight of the experience of individuals, including their actions, as they determine their immediate surroundings within the larger general patterns of the city.

The fictional Wimoa – the lodestar of this study – has been given features that are consistent with the general patterns of the city, as seen from below. But the role of the individual will only be truly appreciated when we take the case of those whose individual situations lead them to actions that do not necessarily fit into larger conceptualizations of the city. The three abstract tales this book began with, sought to capture the possibility of women from very different circumstances being led, by a mix of hope and fear, to actions that brought them to cities that are usually conceptualized as being contrary to needs of their predicament. The first tale, of a woman who feared coming into the big city for the first time, was built around the process of urbanization that remains a critical concern in cities of the global South. As she negotiated her way into the city she would rely on her immediate surroundings, following the Proximity Principle. Her fears about a city she knew little about may have diminished over time as she learnt more about it, but this could have been replaced by new fears including the everyday urban suspense of whether she would be able to achieve her more basic aspiration of educating her child. In responding to this mix of hope and fear, she needed to decide whether to follow the rules that the authorities in the city had laid down, or try to bend them. The actions she chose to carry out would be associated with places in the city, especially her home, her workplace, and those of the various identity groups she associated with. If things went well, the city for her would have been transformed from a huge fearsome entity to a collection of a few places she grew to be comfortable in; places where she had the best conditions to carry out actions prompted by her particular mix of hope and fear.

The preoccupation of the woman software professional in the second abstract tale was to escape her personal fears by moving into a city that was among the more restrictive in the world, especially for women. The city she was escaping from may have placed less restrictions on women as a whole, but the places in it that she was associated with placed severe restraints on what she wanted to do, to a point where she began to fear the actions associated with them. Her home was associated with domestic violence, and her workplace with the failures that resulted from her inability to concentrate, due to the events at home. She knew that the city she was going to live in placed severe restrictions on how she could behave in public places. She was determined to follow the rules she would not have supported in a rational discussion, as a failure to do so would abort her personal escape. The complete surrender to otherwise unacceptable rules gave her greater control over the place of her home. The location of her new home was not in the immediate surroundings of where she grew up, but she ensured its locale reflected her original immediate surroundings. It represented the place she came from in terms of its décor and the television progammes she left on just to hear the sound of her language being spoken. The food she cooked was from the immediate surroundings of her organic space. She conceptualized her home as an autonomous place within a large city with a very different ethos. She had faced a choice between two very different collections of places. In the city she came from, she faced relatively liberal conditions in public places and very restrictive conditions at home. The city she had escaped to offered her freedom in the place of her home with severe restrictions in public places. Like others who valued the personal over the public, she chose the latter collection of places.

The woman in the third abstract tale had allowed hope to overcome her fears. She entered a home in a foreign land that she was certain she would have little control over. The tantrum of disappearance that had been thrown by the man she had married had made her south Asian parents even more determined to ensure she followed the norms he laid out for her. To overcome her apprehensions they had spoken glowingly of the freedom in the United States and the rules that ensured a woman's rights were protected even within her home. They had pointed to the network of relatives she could fall back on in the event of a domestic crisis. She had carefully noted the location of each of the relatives, particularly those in the city she was going to live in. Once in that city she was drawn to the nodes in the network that provided a locale that was closest to what she had had in the immediate surroundings of her parents' home. As she faced the challenges of living with a man she hardly knew, and was not entirely attracted to, she fell back on her preferred nodes in the network of relatives. At times when she faced greater uncertainty she would tend to prefer falling back on relatives who had come to the city even earlier, and hence maintained settings in their home that were closer to those of her grandmother's place in south Asia. They all reassured her that the rules the United States laid out for protecting the rights of women at home did, in fact, work. Her actions increasingly placed a premium on the public over the private in the collection of places that formed the city she now lived in.

The purpose of these three abstract tales, each representing the experience of multiple women from relatively better economic backgrounds than Wimoa, is to emphasize that a city means different things to different people. From these diverse perspectives they carry out actions that they believe are in their interests, even as they face the intended actions of others, as well as things that happen to happen. These actions, even the most personal of them, and the negotiations around them, influence the places they are associated with, and hence the collection of places that form a city. The interventions in this process range from the entirely personal ones of individual women at the lower end of the economic hierarchy, to the policy initiatives of those entrusted with governing the city. The vision that emerges of a city is a snapshot of this process; no matter how comprehensive that snapshot may be, it is but one moment in a continuing process.

Reference

Pani, N., 2009. Resource Cities Across Phases of Globalization: Evidence from Bangalore. *Habitat International*, 33, p. 114–119.

Index

Printed in the United States
by Baker & Taylor Publisher Services